How to Solve Problems

Donald Scarl

How to Solve Problems

For Success in Freshman Physics, Engineering, and Beyond

Second Edition

Dosoris Press, Glen Cove, New York

HOW TO SOLVE PROBLEMS

Published by

Dosoris Press
P. O. Box 148
Glen Cove, NY 11542

Library of Congress Catalog Card Number: 90-81958
ISBN: 0-9622008-2-4

Printed in the United States of America

9 8 7 6 5 4 3 2 1

Contents

③
Methods

④
Describing the Problem

5

Finding
the Solution

⑥
Can't Solve It

⑦
Spreadsheets

Presentation of Results: Reports and Papers

Preface

Most freshman engineering students find that the ability to solve homework and examination problems does not develop easily. Solving problems requires both a knowledge of the material with which the problem deals and a tool kit of methods that can be used to help solve all problems.

Problem solving methods are what this book is about. It presents the methods professional problem solvers use, explains why these methods have evolved, and shows how to make them your own. You can use these problem solving methods now for homework and examinations. In your later professional work they will allow you to define useful problems, solve them, and convince the world that the problems are important and the solutions are correct.

Starting with the pioneering work of Polya, many authors have written self-help books and scholarly books on problem solving. These books address the thinking part of problem solving: how to generalize, specialize, particularize, brainstorm, and so forth. They are especially useful for design problems in which part of the solution may lie in redefining the original problem.

This book, in contrast, emphasizes the simple actions that are taken by professional problem solvers to analyze and solve already defined problems. It explains how to set up and solve problems that you did not think you knew how to solve. It describes the statement, organization, and presentation of scientific and technical material. By teaching problem solving style, it attempts to do for problem solvers what Strunk and White's *Elements of Style* has done for writers.

It is a pleasure to acknowledge Eric Rogers, Jay Orear, and George Wolga, who by their instruction and example helped me to learn how to teach. Hilda Bass, Barbara Cohen, Robert Swart, and Larry Tankersley read the manuscript of the first edition. and made many suggestions. Barbara Cohen, Robert Folk, and Alan Van Heuvelen read and improved the second edition manuscript. Judith Scarl provided reference information for the second edition.

CHAPTER

1

□ □ □ □ □ □   □ □ □ □ □ □

Why Solve
Problems?

How can I pass freshman physics? How can I get a date for tomorrow night? How can I build a user-friendly computer? How can I sell a million of them? People need to solve problems and enjoy solving problems. Problem solving makes our life better and gives us pleasure As we learn methods that make problem solving easier, we can solve harder and more significant problems.

This chapter explains how problem solving, in addition to helping you to pass freshman physics, can help you to learn technical material, to work efficiently, to think clearly, to develop self-confidence, and to convince others to work with you and to support your work.

1

Lifetime learning

Learn in school how to learn by yourself. Science, engineering, and life itself change from year to year, so that most of the learning you will do in life will be done after you leave school and no longer have teachers, homework, tests, and grades to help you learn.. A good way to learn a new subject is to make up problems to solve for yourself. By making up problems you can find the gaps in the way you, and perhaps others, understand a subject. Problem solving is a path to new knowledge and discoveries.

We start out solving problems for others. Your teachers suggest that you do homework problems and insist that you do exam problems. Your employer asks that you solve problems that are part of a group engineering effort. But problem solving is most effective when you begin to solve problems that you make up yourself in order to satisfy your curiosity, to teach yourself a new subject, or to invent or discover something useful to society. Even when doing problems for others, you will do them better and enjoy them more if you pretend that you are doing them for yourself. Doing problems as if for yourself is a good habit.

Workmanship

Human beings have always solved problems; it is part of what makes us human. Problem solving has led to progress, health, security, and comfort. But almost as significant as the solution of a problem is the workmanship used to get that solution. Workmanship, whether in carpentry, writing, or mathematical analysis, is something that can be learned. Part of the pleasure of problem solving is the sense of workmanship and self-esteem that it brings.

Previously solved problems

Although we like to think that concepts are the long-lasting things we learn in school and in life, we also come to depend upon a set of problems that we have learned how to solve. When a new problem comes along we try to make it similar to one of our already solved problems. As a professional scientist or engineer you will often look at your solved and recorded problems to get ideas on how to solve new problems. It is good to get in the habit of documenting solved problems with explicit assumptions and complete algebra so that many years later they will still be usable.

Reflexes

The amount of thinking needed to solve a problem can be reduced by developing good reflexes. The same series of steps can be used to begin every problem. When these steps are automatic, you do not have to think about them and can save your thinking for the parts of the problem that need it. Just as in walking, driving a car, or playing tennis, the more actions that we can make automatic, the higher the level of performance we can achieve. The methods presented in this book will become automatic and raise your ability and performance in problem solving. They apply not only to physics problem solving but to all thinking. Without learning good reflexes you may be able to crawl, but you will not be able to win the race.

Presenting work to others

The homework and test problems that you give to your teachers to evaluate are the public part of your work in school. Professional engineers and scientists give informal talks to their research groups, give talks to technical societies, write quarterly research reports, and write papers for technical journals. The way they analyze, solve, and present problems helps to determine their career success. While you are in school, homework and tests prepare you for these scientific reporting activities.

It is frightening to do work that others will see. "What if I am wrong?" "What if they think that I am not smart?" One of the things that helps us to get over these feelings is practice in handing in work, year after year. The journalist who writes an article every day soon stops worrying about showing his work to others. Handing in homework each week and taking tests every few weeks helps us conquer the fear of having others evaluate our work.

The fear of showing work to others sometimes leads to lack of effort and carelessness. The argument goes something like this: "If I don't try too hard and am sloppy, it shows that I really could do better." Clearly this is a deadly argument and wrong. You will be judged only on the work that you do, not on the work that you *could* do. The right attitude is: "I am going to do my best on every piece of work I do, whether I show it to someone else or not. If my best is still not good enough, I will try to find out whether the problem is mine or someone else's."

Barriers

"I don't think I can do this problem." There is nothing that can stop you as fast as that thought. Do not put that unnecessary barrier in your way. Think "I will be able to do this problem, although it may be harder than I thought and may take more work than I thought." A related barrier is, "I'm afraid I will make a mistake." Fear of mistakes can bring progress to a standstill. Learn how to find and correct mistakes, but have no fear of them. Thinking is difficult. Mental barriers that do not allow young unformed thoughts to pass from the back of your mind into the front must be removed.

"I can't do mathematics (physics, French,...)" Those who seem to be able to do things more easily than you almost always have learned better methods, have developed better work habits, or are simply working harder. In college, talent plays only a tiny role in our ability to learn and succeed. It is true that talent may separate those at the top of their professions from others, but it makes only a minor difference in school.

"This is boring; I would rather be riding my skateboard." It is easy not to be interested in the work we have to do. Surprisingly, it is also easy *to* be interested in the work we have to do. Interests are not in us from birth. They change. They can be changed. A powerful teacher, a clear book, a television program, all can give us a new interest in a subject. You can also do it yourself. "I need to do this, so I *am* interested in it and will try to figure out how to do it well." That idea can't fail.

You can increase your scientific ability, interest, and self-confidence by learning to solve problems well.

Creativity and disorganization

Although the procedures presented in this book are methodical, rational, and powerful, a small number of creative people achieve extraordinary results without them. These people are intuitive, impulsive, and impatient. They have difficulty organizing and explaining their thoughts, their work, and their life. Some good engineers and scientists fall into this category. Methodical procedures slow them down. That is fine for them, but not for you. The great majority of people, and certainly of scientists and engineers, work far better and more productively using methodical procedures. Be one of them.

CHAPTER

2

School

In school you learn facts, procedures, and concepts, and develop good intellectual, professional, and social habits. This book teaches procedures and habits rather than facts or concepts. These procedures and habits will help you to learn, interact with others, and get the pleasures and rewards of scientific and technical accomplishment.

Science and engineering texts

"I understand the material; I just can't do the problems." *Understand* is a slippery word. If at first reading everything seems to make sense, it is easy to say "I understand that." Sometimes making sense is enough, but not often. Understanding means knowing how to use an idea in new ways that were not included in its explanation.

You can only understand an idea by doing problems using it. But what if there are no problems? Make up your own. Develop this habit that allows you to learn and understand on your own. Read an idea. Turn it over in your mind. Make up simple problems. Invent problems testing extreme cases. Solve them. Ask yourself questions. Since no one grades the problems you make up and solve yourself, they can be fun. So much fun that some of the best scientists and engineers appear to want to do nothing else.

Science and engineering books are written differently from other books. While in other books there is sometimes one new idea per paragraph or per chapter (or sometimes per book), in science and engineering books there is often one new idea per sentence. It is not surprising that it takes longer and requires more effort to read these books than others. They are designed not just to be read but to be worked on with pen and paper. When you switch from reading literature to reading physics, it is sometimes hard to remember that your thinking now needs a pen and paper. Work out the mathematical steps that are left out in each explanation and make up and work out problems of your own that lead to real understanding. Making up your own problems after reading each section in a science book lets you understand that section. Make up simple problems, but make up problems.

Learning equations

One aspect of problem solving works backwards. By doing problems you learn to ask the right questions while learning new material.

"I don't know what equations to use to solve this problem." You can prepare yourself for choosing the right equations by classifiing and describing each equation as you learn it. Give the equation a name, describe its importance, classify it, write down the conditions under which it holds, and write a list of the definitions of the symbols used in the equation. This description of equations as you learn them is the key to choosing the right equations to solve a problem later.

An equation's importance

Not all teachers are willing to admit it, but some equations are much more important than others. In introductory physics $F=ma$ takes the prize, with $K=\frac{1}{2}mv^2$ and $p=mv$ close behind. An equation can be important because it is a definition, because it is true under many conditions, because it is simple, because it cannot easily be derived, because it can solve a lot of problems, because it is famous, or because your teacher likes it. While you are learning an equation its importance goes up if your instructor mentions it in lecture, if she assigns a homework problem using it, or if it shows up on a quiz. Learn to be a connoisseur of equations.

Classifying equations

Classify each equation as a definition, a fundamental law of physics, or the solution of a particular problem. If an equation is a definition or a fundamental law, it probably needs to be memorized. If an equation is the solution of a particular problem, maybe it can be derived easily enough, or the problem is specialized enough, that you do not have to memorize it. You can get some idea of the importance of an equation by the amount of space devoted to it in the textbook, by the amount of time spent on it in lecture, and by the number of homework problems that require it.

Go through life knowing the smallest number of equations you can. Spend your time learning the conditions under which an equation holds and the exact meaning of each symbol in the equation, not in memorizing every equation you read. It is better to use a few tools well than many badly.

For each equation, ask yourself: What kind of problem would this equation help me solve? Is it practical to turn this equation around so that I can find one of the quantities on the right hand side in terms of the quantity on the left hand side? In order to find one of the quantities, what other quantities do I need to know?

Special conditions for each equation

An equation holds only under certain conditions. Some equations are much more limited than others. While $F=ma$ is always true, $x=x_0+v_0t+1/2at^2$ is very seldom true. (It is true only for motion with constant acceleration.) Many of the mistakes made in doing problems come from using an equation that doesn't hold under the conditions of the problem. Write down the special conditions when you learn the equation.

Definitions of symbols

Know the exact definition of each quantity that enters an equation. If g is the earth's gravitational constant at the surface of the earth, it will not be the earth's gravitational constant at a satellite that is 20,000 km above the surface. Write down the exact definitions when you learn the equation.

When writing definitions be specific and exact. "The horizontal distance from the starting point" is better than "the distance." "The angle of the initial velocity above the horizontal" is better than "the angle."

Example: How to classify equations

As an example, here is a selection of equations from one chapter of an excellent and popular introductory physics text:

$$v \quad = \frac{d\mathbf{r}}{dt}$$

$$x - x_0 \quad = (v_0 \cos\theta_0)t$$

$$y \quad = (\tan\theta_0)x - \frac{g}{2(v_0\cos\theta_0)^2} x^2$$

$$a \quad = \frac{v^2}{r}$$

$$v_{PA} \quad = \frac{v_{PB} + v_{BA}}{1 + v_{PB}v_{BA}/c^2}$$

After you have read the section in the text that presents one of these equations, classify and describe it on a sheet of paper. When you are finished with the chapter your sheet might look like this:

Definition of velocity

$$\mathbf{v} \quad = \frac{d\mathbf{r}}{dt}$$

Importance: Basic, memorize it.

Class: definition

Conditions: always holds

Symbol definitions: vector velocity of a particle   = $\mathbf{v}$

radius vector from origin to particle $= \mathbf{r}$

Component equations:

$$v_x = \frac{dx}{dt}$$

$$v_y = \frac{dy}{dt}$$

$$v_z = \frac{dz}{dt}$$

Horizontal position of a projectile

$x-x_0$ $= (v_0 \cos\theta_0)t$

Importance: Not very. I can derive it when I need it, from the equations for motion under constant acceleration. Was not mentioned in lecture.

Class: solution of particular problem

Conditions: only for horizontal motion of a free particle with no air resistance

Symbol definitions: horizontal position at $t=0$ $= x_0$

horizontal position at time t $= x$

speed at $t=0$ $= v_0$

angle of velocity vector above horizontal at $t=0$ $= \theta_0$

time at which x is measured $= t$

Trajectory of a projectile

$$y = (\tan\theta_0)\, x - \frac{g}{2(v_0\cos\theta_0)^2}\, x^2$$

Importance: Little. It shows that the trajectory is a parabola. Very complicated and I can derive it from the equations for constant acceleration. Was not mentioned in lecture.

Class: solution of particular problem

Conditions: Vertical position of a particle as a function of its horizontal position. For a particle moving under gravity alone. No air resistance.

Symbol definitions:

vertical position	$= y$
horizontal position	$= x$
size of initial velocity	$= v_0$
angle of initial velocity vector above horizontal	$= \theta_0$
acceleration of gravity at earth's surface	$= g$

Motion in a circle

$$a = \frac{v^2}{r}$$

Importance: Important. Memorize. Use dimensions to check.

Class: special problem, but very general

Conditions:	for motion in a circle with constant speed, but is a good approximation to any motion along a curved path
Symbol definitions:	acceleration toward center	$= a$

	tangential velocity	$= v$
	radius of circle	$= r$

Relativistic addition of velocities

v_{PA}

$$= \frac{v_{PB} + v_{BA}}{1 + v_{PB}v_{BA}/c^2}$$

Importance:	Advanced. Look it up when I need it.
Class:	Special problem. Relative motion of one particle with respect to another when the speeds are close to the speed of light.
Conditions:	Two particles moving along the same line. Works for any velocity but is only necessary when at least one of the velocities is near the velocity of light.
Symbol definitions:	velocity of particle A with respect to my coordinate system	$= v_{PA}$
	velocity of particle B with respect to my coordinate system	$= v_{PB}$
	velocity of particle A with respect to particle B	$= v_{BA}$
	speed of light	$= c$

Homework

Homework is one of the four kinds of work you show to your teachers. The others are tests, laboratory reports, and term papers, When you show your work to someone else, you not only get credit for it, but you are encouraged to work in a way that others can understand. As a bonus, your ideas become clearer to you and. your teacher's opinion of your ability improves.

Homework problems provide practice for professional engineering and science problems. The way you solve homework problems will grow into the way you solve engineering problems. Since school is for practice, your first work in a course does not have to be perfect. Your ability will increase as the semester proceeds. Sometimes you won't notice the improvement since the material also gets harder as the semester proceeds. Notice, though, that last week's homework seems easy to do this week.

The skill you develop doing homework problems carefully will carry over into exams, improving your grades, and into your professional life, increasing your ability and satisfaction.

Exams

Which parts of an exam are the most important? The answers. Yes, but... By concentrating on the answers during the whole exam, your answers will get worse, not better. If you read the problem quickly and interpret it incorrectly, your carefully worked out answer will be wrong. If you work quickly and make a mistake in algebra, your answer will be wrong. If you work quickly and do not write down all your steps, the person who corrects the exam cannot tell what you were doing or how much you know and cannot give you the partial credit you deserve.

The methods suggested in this book seem at first to be impossible to use on exams. How can one work carefully and slowly when there is only one hour to do six problems or even twenty problems? You can. You must. Doing homework problems every week using the proper methods gives enough practice that the same methods can be used efficiently on exams, even when there is much less time than needed. The proper procedures will increase your probability of getting a right answer and increase your average grade. They will assure that you get the maximum amount of partial credit, even when your answer is wrong. They will show the instructor that you understand the material of the course and can work in a professional way.

Sometimes exams look different from other calculations. In particular, multiple-choice problems are dangerous. A problem that happens to have several answers already written down still requires the same calculation on paper as any other problem. If you try to do a multiple-choice problem in your head, you are regressing to old problem-solving habits and decreasing your probability of getting the right answer. Use a sheet of paper or a page of the exam booklet to work out each multiple-choice problem. Find the answer and rejoice when it agrees with one of the choices. Be careful about mistakes, because some of the multiple choice answers will purposely be the ones generated by common mistakes.

You can fall into the same trap in exams as in homework. If you work quickly and sloppily with the subconscious idea that "This is not my best work, I really could do better if I wanted to," you will fool only yourself. Force yourself to do exactly your best work. When that best work is not good enough, ask: "How can I do better next time?"

Efficiency

"Doing problems in my head or just putting down a few steps and getting the answer is quick and efficient." "I don't have enough time to work carefully on exams." "I got all A's in high school but I can't pass college physics." These three thoughts are connected. The method that is most efficient for simple problems is no longer the most efficient for harder problems. One can not expect the method that works well for "If the distance from Lima, Peru to Timbuctoo is 3600 miles and an airplane goes 360 miles per hour ..." to continue to work well for "Find the optimum wing design for maximum lift, minimum drag, minimum weight, minimum moment of inertia, and maximum strength, using graphite fiber composite material." The methods described in this book have been developed by professional engineers, scientists, and designers because they are efficient in getting the right answer, in sharing work with colleagues, in recalling and modifying past work, and in reporting work to others. These professional methods are efficient for homework, exams, and research calculations.

Working in groups

Most modern research and design is done in groups. The ability to solve problems as part of a team is not automatic; it has to be learned. Working with others on homework problems is one way to learn problem-solving teamwork. Each member of the team contributes ideas, each idea is considered and tried, and the best survive. The way that a good team can solve a problem that none of its individual members could solve sometimes appears magical.

In a good team each idea is accepted and appreciated, it is written down, its consequences are explored, it is connected to other ideas, and it is allowed to generate new ideas. When the problem is finally solved, each member of the team understands each of the ideas that went into the solution, whether contributed by herself or by someone else.

However, teamwork has dangers. If ideas are not accepted and encouraged, the most aggressive members of the group will work on their own (not necessarily correct) ideas and the others will sit quietly and listen. The talkers will often go down the wrong path and the listeners will follow them. On the other hand, if the talkers do find the right path, the listeners will accept their ideas, copy down the steps and the result, and believe that they understand how to solve the problem. When a similar problem appears on an exam, where no cooperation is allowed, the listeners will often find that their understanding was not as complete as they thought and that their grade will not be as high it could have been. It is better to work individually than in a group of this kind.

If you cannot solve the homework problems that you need to solve, joining a group can be a good solution. Now your responsibility is to make sure that you question each step the group takes until you understand everything that the group does. This is socially very difficult. Even in a group made up of two people, it is very hard to say, "I don't understand that." It is hard on the ego and it seems to slow down the work. Only the best and most confident scientists and engineers are able to say it easily. In fact, one person saying "I don't understand that," often makes others realize that they don't understand either, and the resulting explanation helps the whole group. "I don't understand that" is one of the most valuable comments in group work.

Professional qualities

Can you develop, while in college, the qualities that will make your professional life more rewarding? Yes, if you know what those qualities are and work toward possessing them. Problem solving can help you reach that goal.

One professiional quality is getting the right answer. Using logical problem-solving methods will help you to get the right answer on tests in school and on professional calculations.

Knowing where to find needed information is a quality that is hard to practice in school. Teachers too often give you everything that is needed. Try to practice finding new information and become familiar with the standard sources of technical information.

Modifying and using past work is a good way to increase your professional efficiency. Doing calculations in a way that lets you come back to them a year later and understand what you did is difficult but valuable.

Sharing your work with others in your group and eventually leading a group can be practiced in college. Learn how to work with a group and how to write up your work so that others in the group can understand it.

Being optimistic is another essential quality. Confidence that you can solve a problem allows you to write carefully and well the parts that you *can* do while not worrying about the parts that you can't do. If you believe that the job can be done and that you can do it, you will be able to convince others to support you and to work with you in a group.

Convincing your supervisors or the scientific community of the worth of what you plan to do is one of the skills that will determine your career success. Both new and continuing projects need the support of others. You must have your plans and calculations clearly written out, and therefore clear in your mind, before you can explain them in a convincing way to others, justify them, and defend them. Presenting a problem and your results to your supervisors, or to the scientific community, is an essential part of technical work. Practice while in school to state the problem and its solution clearly on paper.

CHAPTER

3

□□□□□ **3** □□□□□

Methods

This chapter describes methods that make problem solving easier, that allow clear thinking, and that let you get right answers. These methods are essential when you need to solve problems that you do not know how to solve at first glance. They apply to all parts of the solution of a problem.

Before describing these methods, the next sections are an example of a problem and its solution. The solution has been done using the standard problem-solving methods that we will discuss in this chapter. When we introduce these methods, we will use parts of the solution as illustrations.

Example 1: Distance a baseball can be thrown

An outfielder throws a baseball with a speed of 70 miles an hour at an angle of 45° above the horizontal. How far away will the baseball land?

Solution for Example 1

Distance a baseball can be thrown
D. Scarl
March 3, 1990

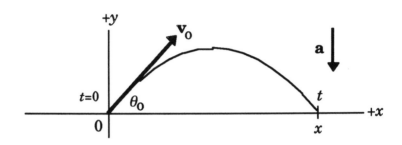

Definitions:

acceleration in x direction	$= a_x$	$= 0$	m/s^2
acceleration in y direction	$= a_y$	$= -g$	m/s^2
earth's gravitational constant	$= g$	$= 9.807$	m/s^2

At $t=0$:

initial x position	$= x_0$	$= 0$	m
initial y position	$= y_0$	$= 0$	m
initial velocity	$= v_0$	$= 70$	mi/hr
initial angle above horizontal	$= \theta_0$	$= 45$	°
x component of initial velocity	$= v_{0x}$		m/s
y component of initial velocity	$= v_{0y}$		m/s

When ball lands:

time at which ball lands	$= t$		s
x position at time t	$= x$		m
y position at time t	$= y$	$= 0$	m

Unit Conversion:

$$v_0 \quad = 70 \ [\text{mi/hr}]$$

$$= (70 \ \text{mi/hr}) \left(\frac{1.61 \ \text{km}}{1 \ \text{mi}}\right) \left(\frac{1000 \ \text{m}}{1 \ \text{km}}\right) \left(\frac{1 \ \text{hr}}{3600 \ \text{s}}\right)$$

$$= 31.31 \ \text{m/s} \ .$$

Geometry:

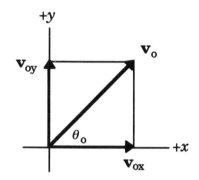

$$v_{0x} \quad = v_0 \cos\theta_0$$

$$= 31.31 \ \cos(45°)$$

$$= 31.31 \ (0.7071)$$

$$= 22.14 \ \text{m/s} \ .$$

$$v_{0y} \quad = v_0 \sin\theta_0$$

$$= 31.31 \ \sin(45°)$$

$$= 31.31(0.7071)$$

$$= 22.14 \ \text{m/s} \ .$$

General Equations:

motion under constant acceleration:

$$x \qquad = x_0 + v_0 t + \frac{1}{2} a t^2 .$$

Particular Equations:

$$x \qquad = x_0 + v_{0x} t + \frac{1}{2} a_x t^2$$

$$ \qquad = 0 + v_{0x} t + 0 .$$

$$y \qquad = y_0 + v_{0y} t + \frac{1}{2} a_y t^2$$

$$0 \qquad = 0 + v_{0y} t + \frac{1}{2} a_y t^2 .$$

Calculations:

Solve y equation for the time t at which the ball lands

$$0 \qquad = v_{0y} t + \frac{1}{2} a_y t^2$$

$$\frac{1}{2} a_y t^2 \qquad = -v_{0y} t .$$

$$\frac{1}{2} a_y t \qquad = -v_{0y}$$

(t has cancelled, so t=0 seems to be a solution, too)

$$t \qquad = \frac{-v_{0y}}{\frac{1}{2} a_y}$$

$$ \qquad = \frac{-22.14 \text{ m/s}}{\frac{1}{2}(-9.807 \text{ m/s}^2)}$$

$$ \qquad = 4.515 \text{sec} .$$

Solve x equation for x

$$x \quad = v_{0x}t$$

$$= v_{0x} \left(\frac{-v_{0y}}{\frac{1}{2}a_y} \right)$$

$$= \frac{-2v_{0x}\, v_{0y}}{a_y}$$

$$= \frac{-2\,(22.14\text{m/s})(22.14\text{m/s})}{-9.807\text{m/s}^2}$$

$$= 100 \text{ m }.$$

Result:

$\boxed{\text{distance thrown} = x = 100 \text{ meters}}$.

Divide into parts

Dividing the solution into parts is the first step in solving any problem. It is useful for problems at all levels, and is essential for large problems. Professional programmers know that one of the hardest tasks in writing a large program is dividing the program into parts that can be worked on separately and recombined smoothly. Divide the solution into the smallest possible parts and work on each part separately.

The parts into which most introductory science and engineering problem solutions can be divided are:

> Heading
> Labelled drawing
> Symbol definitions
> Data
> Preliminary equations
> Science equations
> Calculation
> Results.

Chapters 4 and 5 will describe what each of these parts contains.

A second way to divide a problem into parts is by dividing it in space and time. If it involves several different masses, try to describe the conditions for each mass separately. If it is a circuit with many operational amplifiers, first draw and analyze each amplifier separately. If the events in the problem take place at two or more different times, describe separately what is happening at the first time, the second time, and so forth, and then write the equations that connect the events at the different times.

Do the parts separately

First divide a problem into parts. Then force yourself to think about the parts separately. A common mistake is to try to solve the whole problem at once. As you start to work on a problem, it is tempting to try to figure out the answer or to worry about what equations apply and how to solve them. These are not the parts of the problem to think about at the beginning. First do the simple and automatic steps that begin the solution of every problem: write a heading, draw a labelled diagram, define your symbols. None of these steps requires knowledge of how you will eventually solve the problem.

Professional problem solvers learn not to worry about the parts of the solution that they cannot do, while they work on the parts they can do. If a problem is difficult, describing it clearly and dividing it into parts is a useful contribution, even if some of the parts cannot be done by you or by anyone else. Describing what needs to be done is a first step toward getting it done.

Dividing a solution into parts is useful on tests. If you sit and worry about the part of the problem you can't do, you will get no credit for writing down the parts you can do. After writing all of the automatic steps such as the definition of symbols and the geometry equations, you may understand the problem well enough to be able to solve it, even though you didn't think you could. Divide the solution into parts and do the parts you can do.

Label each part

Begin each part of the solution with a short title saying what you are starting to do:

> **Distance a baseball can be thrown**
> :
>
> **Definitions**
> :
>
> **Unit conversion**
> :
>
> **Geometry**
> :
>
> **General equations:**
> :

Writing a title for each part of the solution helps in thinking about that part of the solution and helps even more when you come back to the problem later or show the problem to someone else. On tests, writing titles helps your instructor understand what you are doing and leads to a higher grade. These titles are part of the documentation of the problem; they make clear what you are doing on each line. Algebra is a very condensed way of expressing ideas. Break up the algebra with titles describing what you are about to do.

A special case of this rule is to write a one-line description of the problem at the beginning of your work:

> **Distance a baseball can be thrown**
> :

This helps those absent-minded professors who do not remember exactly what problem they assigned, and helps *you* remember exactly what the problem was when you look back at a problem that you solved years ago.

Work down the page

The graphic arrangement of your work can make it easier to read. Work down the page, one thing under the next. Jumping to a second column or writing things side by side interferes with being able to see the order of the solution easily. Writing up the side of the page or squeezing the last few lines in at the end of a page is not a good idea. Start a new numbered page. Think of the person reading or grading your paper. Can she understand the order of your solution and give you credit for all the parts that you have done?

The right hand side of an equation can go across the page with successive equal signs but it is better to start a new line for each new equal sign. When you go to the next line with another step of the same equation, line up the equal sign with the one on the line above:

Solve x equation for x:

$$x \qquad = v_{0x}t$$

$$= v_{0x}\left(\frac{-v_{0y}}{\frac{1}{2}a_y}\right)$$

$$= \frac{-2v_{0x}\,v_{0y}}{a_y}$$

$$= \frac{-2\,(22.14\text{m/s})(22.14\text{m/s})}{-9.807\text{m/s}^2}$$

$$= 100 \text{ m}\ .$$

Write on one side of the paper. Compared with the cost of your education, paper is inexpensive. No professional calculator has the patience to look on the back of pages to find the part of the solution he is looking for. The only exception to this is in test booklets or notebooks where writing on both sides is necessary and expected.

Use horizontal lines to separate parts of the solution or to show that you have stopped working on one equation and have begun working on another.

Use solid lines, dashed lines, dotted lines, squiggly lines, double lines, or whatever you like to show what is going on. Use them where we use **boldface** in this book or wherever you want. The more aids to the eye you use, the easier it is to understand your work.

Calculations:

Solve y equation for the time t at which the ball lands

$$0 \qquad = v_{0y}t + \tfrac{1}{2}a_y t^2$$
$$\vdots$$

Use boxes for answers. Put the e of the variable and its unit in the box. Make it easy for the person reading your solution (it could be yourself) to find what they need:

Result:

distance thrown = x = 100 meters .

Write clearly

Your handwriting reveals your attitude. Slow down and write beautifully. Here again, teachers can set a bad example. They write quickly on the blackboard so that they can cover more material in an hour. When you are solving a problem there is no need to write quickly. The time taken in writing is a small fraction of the time spent working on the problem. If the way you always write a symbol allows it be confused with another symbol, change the way you write that symbol. Write a zero before naked decimal points to make sure the decimal point doesn't get overlooked: x = 0.17 m. If you accidentally write a symbol or number unclearly, cross it out and write it again clearly. Each symbol in a long calculation will be written many times. If it is unclear even once, it can lead to confusion and mistakes. You are the one who will suffer the most from writing that is not carefully done. Writing sloppily is like stabbing yourself in the foot. A year later when you cannot understand what you have written, it will hurt a lot.

Make everything explicit

Write everything down. So easy to say, so hard to do. Facts that are kept in your head instead of being written down make it much harder to think about the problem. Facts that are not written down cannot be checked for mistakes. Facts that are not written down cannot be used or even understood when you come back to the problem in a few years. Facts that are not written down will generate no credit on an exam. If something is part of the problem, write it down:

Definitions:

acceleration in x direction	$= a_x$	$= 0$	m/s^2
acceleration in y direction	$= a_y$	$= -g$	m/s^2
earth's gravitational constant	$= g$	$= 9.807$	m/s^2

At $t=0$:

initial x position	$= x_0$	$= 0$	m
initial y position	$= y_0$	$= 0$	m
initial velocity	$= v_0$	$= 70$	mi/hr
initial angle above horizontal	$= \theta_0$	$= 45$	°
x component of initial velocity	$= v_{0x}$		m/s
y component of initial velocity	$= v_{0y}$		m/s

When ball lands:

time at which ball lands	$= t$		s
x position at time t	$= x$		m
y position at time t	$= y$	$= 0$	m

Avoid thinking

Save your thinking for when it is necessary. Learn to describe a problem, write the data equations, draw the picture, do the geometry, and write the general equations automatically without thinking. Make these actions into reflexes. Using the same methods for every problem avoids thinking, sometimes at the cost of going up a few wrong paths, but saving work in the long run. The bicycle racer who doesn't have to think about balancing can think about pacing and strategy instead.

Use symbols

Although numbers are given in the statement of a problem and numbers are needed for the answer, do not use numbers in setting up the solution or in writing the equations. It is necessary to make up a e and a symbol for every quantity that enters into the problem. First solve the problem with symbols. Use algebra to get the symbol for the answer by itself on the left hand side of an equation that has only symbols on the right hand side. Then put in the numbers and find the answer.

Using symbols for quantities is very hard to do at first. The equations look more complicated with symbols than they do with numbers. With symbols the equations look like algebra, while with numbers they look like only arithmetic. Don't be frightened; algebra works at least as well as arithmetic once you are comfortable with the rules. Of course, it is difficult to do very much in your head with symbols. But the goal is *not* to do anything in your head.

By working without numbers you can solve the problem in general. If the input numbers change, as they often do in a professional problem, all you have to do is reenter them in the algebraic equation for the answer and get a new result. Often the result of a professional problem is presented as a graph of how the answer depends on one or more of the initial conditions. The graph can be calculated easily if you have the answer as an algebraic expression, so that by substituting values of the independent variable you can calculate corresponding values of the dependent variable.

If you use numbers during the solution of a problem, you have to go through the whole solution changing numbers when the initial conditions change. Programmers know not to use numbers in the body of a computer program or spreadsheet. Numbers are hard to find once they are absorbed into a calculation. They belong only at the beginning:

Definitions:

acceleration in x direction	$= a_x$	$= 0$	m/s^2
acceleration in y direction	$= a_y$	$= -g$	m/s^2
earth's gravitational constant	$= g$	$= 9.807$	m/s^2

$\vdots$

and at the end:
Solve x equation for x

$$x \quad = v_{0x}t$$

$$= v_{0x}\left(\frac{-v_{0y}}{\frac{1}{2}a_y}\right)$$

$$= \frac{-2v_{0x}\,v_{0y}}{a_y}$$

$$= \frac{-2\,(22.14\text{m/s})(22.14\text{m/s})}{-9.807\text{m/s}^2}$$

$$= 100 \text{ m .}$$

Another advantage to avoiding numbers and forcing yourself to calculate with symbols is that the dependence of the result on the initial conditions can be seen and checked:

$$x \quad = v_{0x}t$$

$$= v_{0x}\left(\frac{-v_{0y}}{\frac{1}{2}a_y}\right)$$

$$= \frac{-2\,v_{0x}\,v_{0y}}{a_y}\;.$$

Does this equation for the distance travelled show that the distance increases as the speed increases? Does it show that the distance decreases as the downward acceleration increases? These questions are good checks that the algebra was done correctly. They also let you understand the form of the result and its sensitivity to each of the initial conditions. If you are lucky, some of the quantities may even cancel, telling you that they were not an essential part of the problem, or at least saving time and errors in the final calculation with numbers.

By keeping symbols through the working of the solution instead of putting in numbers, the unit can be checked in the result. If the unit on the right hand side of the result equation does not agree with that on the left, there is a mistake somewhere. When you have made a mistake by leaving out a necessary factor, the ratio of the unit on the right side to the unit on the left can tell you the unit of the missing factor, helping you to find it. Checking units is a powerful way to catch errors.

On the other hand, if you cannot understand a problem or find its solution, putting in numbers can help you to get started or help you to solve equations. Chapter 6 ecplains how this works..

Zero

There is an exception to the rule about saving numbers for the end of the problem   After you have written one of the full famous general equations that will help you to solve a problem and have written a particular case of that equation using the variables you have defined for your problem, it is convenient to substitute zero for quantities that are zero:

Particular Equations:

$$x \qquad = x_0 + v_{0x}t + \frac{1}{2}a_x t^2$$

$$ \qquad = 0 + v_{0x}t + 0 \ .$$

$$y \qquad = y_0 + v_{0y}t + \frac{1}{2}a_y t^2$$

$$0 \qquad = 0 \ + v_{0y}t + \frac{1}{2}a_y t^2 \ .$$

By putting in zeros as soon as possible (never in the original writing down of the famous equations) equations can be shortened and the problem can be solved more easily. The price that you pay for this is, if the initial conditions change so that a variable is no longer zero, you need to go back and redo the entire solution. Even if this happens, you then have your first simplified version with which to compare the more complicated result.

Do one step at a time

Do algebra one step at a time.

When doing algebra, write down every step, one step at a time. The step that you do not write down is the one that contains the mistake.

There are, unfortunately, too many reasons why writing one step at a time does not come naturally. Almost all teachers who write equations on the blackboard do several steps at a time. All textbooks and technical papers do several steps at a time. They do this to save time and paper. You can be sure that these same teachers and authors worked out the solution one step at a time before they wrote on the blackboard or in their book. The way a solution is finally presented in a lecture or book is not good to use as a model when doing your own problem solving.

The reason for doing one step at a time is to avoid mistakes. If the probability of doing one step correctly is 60%, the probability of doing two steps correctly is 36%, and the probability of doing three steps at a time correctly is 22%. If each step is not written down separately, each step cannot be checked for mistakes. A single algebraic mistake makes all of the results that follow it wrong. Mistakes near the top of a solution lead to even more trouble than those near the bottom. Mistakes are expensive. They waste time and generate frustration. Work in a way that minimizes mistakes.

One step at a time.

Use ratios

A ratio (the quotient of one quantity divided by another) can often be found even when the quantities themselves cannot. The ratio of two quantities with the same units is "dimensionless" and is convenient for calculation. A dimensionless ratio gives the value of the top quantity measured in units of the bottom quantity. It also tells what fraction the top quantity is of the bottom quantity.

Quantities like efficiencies and power ratios measured in db (decibels) are ratios or logarithms of ratios by definition. Others like the Reynolds number in fluid dynamics and the Strehl ratio in optics are useful dimensionless numbers that describe physical conditions. New laws can be discovered by noticing that an experimental result depends only on the ratio of two quantities or that the ratio of two quantities is always a constant.

Ratios are helpful in setting up and solving problems. Use them whenever you can.

Do the assigned problem

In a homework or test problem make sure to do the problem that is written down and to find the solution that is asked for. Do not do more or less than the problem asks. If the instructions ask you to set up the problem but not to solve it, do not solve it. If the instructions ask you to use an equation but not to derive it, do not derive it.

A problem statement that asks: "Get the result in terms of y and z" asks you to solve for an equation with the result alone on the left hand side and nothing but numbers, constants, and y and z, or functions of y and z (y^3, $\sin z$, etc.) on the right hand side. If other quantities appear on the right hand side of the equation for the result, find their dependence on y and z and substitute into the result equation:

Distance a baseball can be thrown
Find the distance a baseball can be thrown, in terms of
the speed of the throw, its angle above the horizontal
and the earth's gravitational constant:

$$x \quad = v_{0x}t$$

$$= v_{0x}\left(\frac{-v_{0y}}{\frac{1}{2}a_y}\right)$$

$$= \frac{-2v_{0x}\,v_{0y}}{a_y}$$

$$\boxed{x \quad = 2\,v_0{}^2\,\frac{\cos\theta\,\sin\theta}{g}}$$

CHAPTER

4

☐☐☐☐☐☐ 4 ☐☐☐☐☐

Describing
the Problem

Now that we have seen the methods that are used for all parts
of the solution of a problem, we can discuss the separate parts of
the solution. This chapter explains the first part of the solution:
setting up the problem by drawing a diagram and defining
symbols.

The next two sections are an example of a problem and its
solution. The parts of the solution will be used to illustrate the
material in this chapter and in Chapter 5.

Example 2: Ultralight plane takeoff

An ultralight human-powered airplane and its pilot
weigh 200 pounds. The pilot's pedalling drives a
propeller that produces a forward force on the plane. The
plane's motion through the air produces a force on its
wings equal to a constant times the square of the
plane's velocity. The direction of this air-flow force is 10°
behind straight up, and the constant is 25 N/(m/s)^2. On

takeoff, the pilot accelerates the plane along the
runway with a constant acceleration of 0.6 m/s². The
plane lifts off the ground when the upward component of
the air-flow force on its wings equals its weight. How
long a runway will the plane need to just lift off the
ground?

Solution for example 2

Ultralight Plane Takeoff
D. Scarl
March 4, 1990

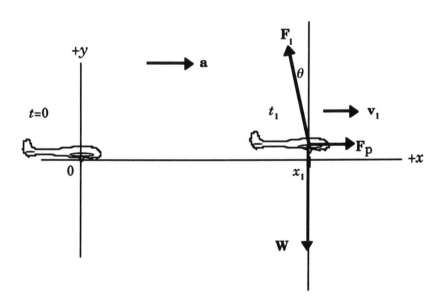

Definitions and Data:

weight of plane and pilot	$= W$	$= 200$	lb
acceleration along runway	$= a_x$	$= 0.6$	m/s^2
upward acceleration	$= a_y$	$= 0$	
air-flow force constant	$= K$	$= 25$	$\dfrac{N}{(m/s)^2}$
angle of force from vertical	$= \theta$	$= 10$	°

At beginning of runway:

time at beginning	$= t_0$	$= 0$	s
distance along runway	$= x_0$	$= 0$	m
velocity	$= v_0$	$= 0$	m/s

At takeoff:

time at takeoff	$= t_1$		s
distance along runway	$= x_1$		m
velocity	$= v_1$		m/s
force on plane from air flow	$= F_1$		N
upward air-flow force	$= F_y$		N
backward air-flow force	$= F_x$		N
force on plane from propeller	$= F_p$		N

Unit Conversion:

$$W \quad = 200 \text{ lb} \qquad = 200 \text{ lb } (4.448 \; \frac{\text{Newtons}}{\text{lb}})$$

$$= 889.6 \text{ Newtons}$$

Geometry:

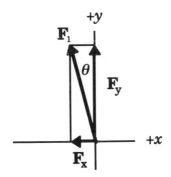

$$F_y = F\cos\theta$$

$$F_x = F\sin\theta$$

General equations:

equations for constant acceleration

$$a = \text{constant}$$

$$v = v_0 + at$$

$$x = x_0 + v_0 t + \frac{1}{2} at^2$$

connection between airflow force and speed

$$F_1 = Kv^2$$

condition when plane just leaves ground,

$$F_y = W$$

Particular equations and vector components:

To find v_1, the velocity at takeoff

$$F_1 = Kv_1^2$$

$$v_1^2 = \frac{F_1}{K}$$

---

from the geometry equations:

$$F_1 = \frac{F_y}{\cos\theta}$$

$$v_1^2 = \frac{F_1}{K}$$

$$= \frac{F_y}{K\cos\theta}$$

$$v_1 = \sqrt{\frac{F_y}{K\cos\theta}}$$

at takeoff, the upward component of the airflow force equals the weight

$$F_y = W$$

$$v_1 = \sqrt{\frac{F_y}{K\cos\theta}}$$

$$v_1 = \sqrt{\frac{W}{K\cos\theta}}$$

$$= \sqrt{\frac{889.6 \text{ N}}{(25\text{N/m}^2/\text{s}^2)\,(\cos 10°)}}$$

$$= \sqrt{\frac{889.6 \text{ N}}{(25\text{N/m}^2/\text{s}^2)\,(0.985)}}$$

$$= \sqrt{36.13 \text{ m}^2/\text{s}^2}$$

$v_1 \qquad = 6.01 \text{ m/s}$

To find t_1, the time to reach takeoff speed

For motion under constant acceleration:

$v \qquad = v_0 + at$

$v_1 \qquad = 0 + a_x t_1$

$t_1 \qquad = \dfrac{v_1}{a_x}$

$\qquad \quad = \dfrac{6.01 \text{m/s}}{0.6 \text{m/s}^2}$

$\qquad \quad = 10.02 \text{ sec}$

To find x_1, the position at time t_1

For motion under constant acceleration:

$x \qquad = x_0 + v_0 t + \frac{1}{2} a t^2$

$x_1 \qquad = 0 + 0 + \frac{1}{2} a_x t_1^2$

$\qquad \quad = \frac{1}{2} a_x t_1^2$

$\qquad \quad = \frac{1}{2}(0.6 \text{m/s}^2)(10.02 \text{s})^2 \quad = 30.1 \text{ m}$

$\boxed{\text{length of runway} = x_1 = 30.1 \text{ meters}}$

Check by getting algebraic solution:

$$x_1 = \frac{1}{2}a_x t_1^2$$

$$= \frac{1}{2} a_x \left(\frac{v_1}{a_x}\right)^2$$

$$= \frac{1}{2}\frac{1}{a_x} (v_1^2)$$

$$= \frac{1}{2}\frac{1}{a_x} \left(\frac{W}{K\cos\theta}\right)$$

$$= \frac{889.6 \text{ N}}{2 \, (25\text{N-s}^2/\text{m}^2)(0.6\text{m/s}^2)(\cos 10°)}$$

$$= \frac{889.6 \text{ N}}{2 \, (25\text{N-s}^2/\text{m}^2)(0.6\text{m/s}^2)(0.985)}$$

$$= 30.1 \text{ m}$$

length of runway = x_1 = 30.1 meters

Define the problem

The first part of the attack on a problem is to get *all* of the information about the problem on paper. This holds for homework problems, test problems, and especially for professional problems in which the problem itself usually is not clear at first. Put a complete list of the facts of the problem on paper so that you can close the book, or not look at the test paper again, for that particular problem. Do not copy the problem itself from the book or assignment sheet. Instead, write the facts of the problem as equations and diagrams.

Putting the problem on paper forces you to understand each part of the problem, it puts everything together in one place, it allows someone else to understand exactly what the problem was, and it begins your thinking about the problem in an easy way without the terror of worrying about the answer or what equations to use. It is a good example of working instead of thinking. (It also shows the person grading a test that you understood the statement of the problem.)

Sometimes problems are not stated clearly in homework assignments, on tests, and especially in professional science and engineering. By writing the facts of the problem as you understand them, you are telling the person reading the solution what assumptions you made for those parts of the problem that were ambiguous. If you write your interpretation of the problem and make your assumptions clear, you can get credit for solving the problem using those assumptions.

Once the complete problem is on paper, go back, read the problem again, and check that you have every fact down correctly. Are those units centimeters or meters? Did you see that the motion started from rest, and do you have $v_0=0$ on your paper? Just one wrong fact at this point will make the whole solution wrong. If you have a question about the statement of a problem on a test or on a homework assignment, describe the problem on paper and then ask the instructor your question.

Write a heading

Title

Write a one-line title describing the problem:

Ultralight Plane Takeoff

This is a first easy step in thinking about the problem and also lets you remember what the problem was about when you look at it a month later. The problem number as given in a book or assignment sheet is needed and should be written down, but does not give much information about the problem and does not take the place of a title. Professional problems have no numbers but they do have titles.

Name and date

Every piece of written work that you do requires your name and the date as part of its heading: your name, so that when you copy the work and show it to someone else they will know who the author was, the date, so that after you have solved a problem several times with increasingly good results, you will know which is the latest version.

Draw a diagram

Start with a diagram:

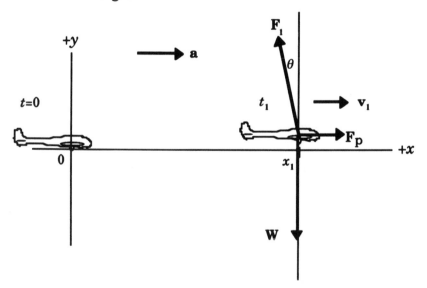

Most people think in pictures and find a diagram a help even in problems that have no obvious geometrical part. Venn diagrams in logic, organizational charts in management, and even outlines of written papers are diagrams that help one to think.

Diagram or picture

A picture can be artistic, subjective, and abstract in order to suggest objects or emotions. A diagram, on the othe hand, shows, realisticallly and to scale, the objects that are part of the problem, together with abstract symbols such as axes and vectors.

Sometimes the problem needs several diagrams. Use one to show the initial conditions, another to show the final conditions, one on which to draw the trigonometric relationships, and perhaps one to display the answer. In problems that have things happening at different times, a diagram, like a snapshot, at each of the times the problem talks about is helpful.

Draw your diagrams large enough that all angles and distances are clear, and are roughly in proportion to their actual values.

When not enough is known about the answer to draw a correct diagram, just make reasonable assumptions and draw a diagram anyway. If the diagram is not right, the answer will tell you so, by giving a negative number, for instance.

Try to draw a diagram that is not a special case. If the problem is about a triangle, draw a general triangle, not one that includes a right angle or one that has equal sides. A diagram of a pendulum at a general place in its swing is more useful than one with the pendulum at the top or bottom of its swing:

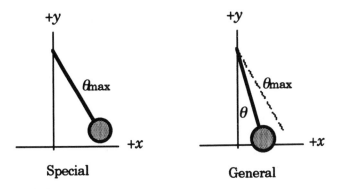

On the diagram on the right both the angle of the pendulum θ and the special angle θ_{max} can be shown. On the diagram on the left, θ cannot be shown.

Axes and vectors

Most diagrams need a set of axes. Draw x and y axes and label the $+x$ and $+y$ directions and the zero position:

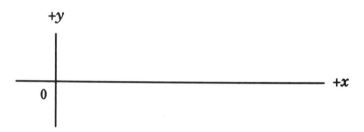

On these axes any vector component pointing in the $+x$ or $+y$ direction is positive. A vector component pointing in the $-x$ or $-y$ direction is negative

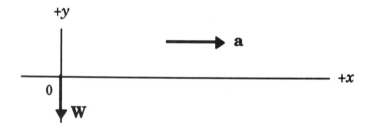

. Although the x and y axes have to be perpendicular, they do not have to be horizontal and vertical. When the problem involves known vectors, the wisest choice of axis directions is the one that aligns the axes with as many of the given vectors as possible. For instance, in a simple inclined plane problem axes along the plane and perpendicular to it align along an axis all of the forces except the weight.

In diagrams showing force vectors as well as velocity and acceleration vectors, attach the force vectors to the objects they act on. Draw the velocity and acceleration vectors near the object they describe but not touching, so that they are not confused with the force vectors.

Name the variables

Each number that enters into the problem needs to be given a name: each constant, each variable, each initial condition, and each result. While in a spreadsheet or computer program the name can be several letters that describe the quantity (INITVEL), in most solutions the name is a symbol, a letter decorated with subscripts (v_0). Although a quantity is given as a *number* in the statement of the problem, give it a *symbol* when you work on the solution, saving the numbers until the last step.

Symbol definitions

Make a list of *all* of the quantities that enter the problem:

Definitions and Data:

weight of plane and pilot	$= W$	$= 200$	lb
acceleration along runway	$= a_x$	$= 0.6$	m/s^2
upward acceleration	$= a_y$	$= 0$	
air-flow force constant	$= K$	$= 25$	$\dfrac{N}{(m/s)^2}$
angle of force from vertical	$= \theta$	$= 10$	°
At beginning of runway:			
time at beginning	$= t_0$	$= 0$	s
distance along runway	$= x_0$	$= 0$	m
velocity	$= v_0$	$= 0$	m/s
At takeoff:			
time at takeoff	$= t_1$		s
distance along runway	$= x_1$		m
velocity	$= v_1$		m/s
force on plane from air flow	$= F_1$		N
upward air-flow force	$= F_y$		N

backward air-flow force $\qquad = F_x \qquad$ N

force on plane from propeller $\qquad = F_p \qquad$ N

Define the given quantities or initial conditions, useful intermediate quantities that you can use to help solve the problem, and the unknown quantities that will be the results. The definitions can be organized in any way you want. One useful way is first to define those quantities that remain constant, then those quantities that describe what is happening at t=0, then those that describe what is happening at t_1, and so on. At each time you may need to define angles, positions, velocities, accelerations, forces, momenta, energies, and other quantities. Start with the simplest and build up to the more complicated.

Quantities that are zero have equal rights with quantities that have other values. When the initial velocity is zero, write it down. If everything you do in the solution of the problem depends on the acceleration being zero, it makes sense to write, at the beginning of the data equations:

acceleration $\qquad = a \quad = 0 \quad$ m/s^2

For some reason, perhaps because zero length vectors are difficult to draw on the diagram describing the problem, zero quantities are easy to forget. Write them down so that you have a symbol for them and to remind yourself and others that they are part of the statement of the problem or are needed to help solve the problem.

Write the unit for each quantity. It is amazing that just writing the unit for a mysterious quantity makes it less mysterious and more understandable. It also avoids mistakes when the quantity is given in centimeters and you are working in meters.

Label the parts of the diagram with the symbols that you have made up for each quantity:

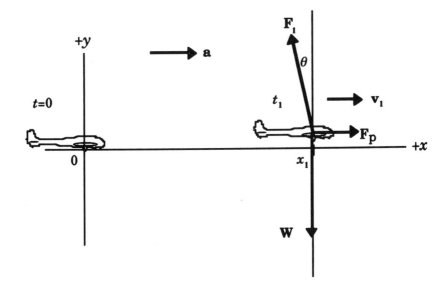

Do not put numbers on the diagram for the same reason that you do not put numbers into the beginning of the calculation. You may want to change the numbers later without redoing the diagram or restarting the solution. Numbers belong only in the definition of each quantity in your list of data equations, in the last few equations of the calculation, and in the final result.

Choosing names

If the names of the quantities you need are given in the problem or are famous, choosing names is easy. If the names are not given, or if only the values of the quantities are given, make up a name for each quantity that will be needed to set up the solution. Here are some generally accepted rules for naming quantities:

x, y, z	distances across, over, and up
i, j, k, l, m, n	integers
a, b, c	constants
$\alpha, \beta, \gamma, \theta$	angles

Subscripts

Symbols like v for velocity need to be decorated with subscripts when there are several different velocities in the problem. Subscripts describe *which* velocity you are talking about.

Subscripts that are numbers usually indicate that the quantity is a constant:

> initial velocity = v_0 = 0 m/s
> (Although v may change with time, v_0 is a constant.)

The best way to label times is with numerical subscripts, so that all the variables at a given time can have the same subscript. x_1, v_1, and p_1 are the position, velocity, and momentum at time t_1. They are constants and each has only a single value. x, v, and p are the position, velocity, and momentum at any time t. They are variables and have different values for every value of t.

Subscripts that are small letters usually indicate that the quantity is a variable:

> velocity in the x direction = v_x
> (It changes with time. Notice that v_{x0} (or v_{0x}) is the
> initial velocity in the x direction and is a constant.)

Subscripts that are capital letters usually describe different objects. M_A and M_B are the masses of object A and object B and are constants, while $\mathbf{p}_A$ is the momentum of object A and is a variable.

Data equations

Sometimes the definition of a quantity is all you have at the beginning. However, if a *number* is given in the statement of the problem, it can be included in the definition. When the definition includes a number, it becomes a data equation.

Definitions and Data:

weight of plane and pilot	= W	= 200	lb
acceleration along runway	= a_x	= 0.6	m/s^2

air-flow force constant $\qquad = K \qquad = 25 \qquad \dfrac{N}{(m/s)^2}$

angle of force from vertical $\qquad = \theta \qquad = 10 \qquad °$

$\vdots$

The data equations turn the written statement of the problem into a set of equations. Read each phrase of the problem and write down the equation it generates. The sentence, "A ball is dropped from a height of 7 meters," generates these two data equations:

initial velocity $\qquad = v_0 \qquad = 0 \qquad$ m
initial height $\qquad\quad = y_0 \qquad = 7 \qquad$ m.

Watch especially for these phrases and the equations they generate:

"starts from rest":
initial velocity $\qquad\qquad = v_0 \qquad = 0 \qquad$ m/s

"moves with constant velocity":
acceleration $\qquad\qquad\quad = a \qquad = 0 \qquad$ m/s^2

"two equal masses" :
mass of body 1 $\qquad\qquad = m_1 \qquad\qquad$ kg
mass of body 2 $\qquad\qquad = m_2 \qquad = m_1 \qquad$ kg

Sometimes a definition and a data equation look almost alike:

Definition:
$F = F_1 + F_2$
Data Equation:
$F = 0.$

The definition equation $F=F_1+F_2$ defines the total force F as the sum of force F_1 and force F_2. The data equation says that for some reason the total force F is zero. It is tempting to combine these two different equations into one equation: $F_1+F_2=0$. But now you have no symbol for the total force, and it may not be at all obvious why the sum of F_1 and F_2 is zero. Be careful to separate definition equations and data equations and to write down both .

In the data equations and in the calculation every equation has to have a left hand side. Writing down ma by itself doesn't mean anything at all.

Check

After drawing a diagram and writing down all the symbol definitions and data equations that the statement of the problem generates, read the problem again and check that every phrase has its own equation. Check your definition of variables against the problem statement definition of variables. Check that you have named a quantity that corresponds to the answer, even though you have no idea what the answer is.

Preliminary equations

There are two kinds of equations needed to complete the solution to a problem. The first equations are those that prepare the data and get them ready for the science equations that are the second part. The preparations that make up the first part include changing units, looking up the values of constants, and calculating components of vectors and other trigonometric ratios.

These preliminary calculations often need their own new diagrams. The original diagrams that defined the problem should not be complicated by these new calculations that are not part of your statement of the problem.

Units

There is no longer any choice of units. All of your courses will use the meters, kilograms, seconds (SI) system. If you live in the United States and become a civil or automotive engineer, an architect, a machinist, or a carpenter, you can learn to use English units on the job.

If parts of the problem are given in non-SI units, change to SI. Change units in a way that can be checked later. Write the quantity with its original unit, an equals sign, the original quantity again, the multiplication factor that changes the unit, an equals sign, and the quantity with the new unit:

$$W \quad = 200 \text{ pounds}$$
$$= 200 \text{ pounds} \left(\frac{4.448 \text{ newtons}}{1 \text{ pound}} \right)$$

$$= 889.6 \text{ newtons}$$

There are, unfortunately, some exceptions to this rule. If the problem is given entirely in non-SI units, and the answer is acceptable in non-SI units, solve the problem with the given units. If the problem is stated in SI units but you are asked for the answer in non-SI units, calculate the solution in SI units and then convert the answer at the very end. These cases still arise on tests, but less and less often in scientific and engineering practice.

Constants

Write down the value of constants that are needed to solve the problem. Sometimes these will be given in the statement of the problem and sometimes you will have to look them up. For each constant you look up write a reference to the book or paper from which you got the value of the constant. If the constant is a rare one and hard to find, you will appreciate the reference when you come back to it later.

earth's gravitational constant = g

$$g \quad = 980.665 \text{ cm/s}^2$$
$$= 980.665 \text{ cm/s}^2 \left(\frac{1 \text{ m}}{100 \text{ cm}} \right)$$

$$= 9.807 \text{ m/s}^2$$

From C. W. Allen *Astrophysical Quantities*, pg. 109, Athlone Press, London 1955

Trigonometry and vector components

Geometry is the cause of much of the complication of science and engineering problems. Especially in three dimensions, descriptions of points, directions, and fields become geometrically difficult. By doing the geometrical calculations first you can separate the geometry from the science and simplify the solution.

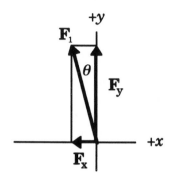

Geometry:

F_y $= F\cos\theta$

F_x $= F\sin\theta$

CHAPTER

□□□□□ **5** □□□□□

Finding the Solution

Science equations

Now that the problem is exactly stated on your paper, the drawing is clear and beautiful, all of the data equations are written, units have been converted, auxiliary drawings have been made, constants have been found, and the trigonometry has been done, you are ready to think about the physics. Notice that up to now you have not had to think. All of the preparation you have done is the same for every problem and will soon become automatic.

Instead of beginning at the beginning, amateur problem solvers often begin at this point and then wonder why problem solving is so difficult and their exam scores are so low. By doing all of the standard preparation you have been thinking about the problem in the back of your mind for several minutes. You have all of the necessary information about the problem on paper. On an exam you have impressed the test grader that you know something about the problem and you have already gotten about one-quarter credit for it, all without knowing how to do the problem.

Choosing the right equation

Now some thinking helps. Not the kind of thinking that tries to solve the problem in one bite, but the thinking that asks, "What is going on here? Is the velocity constant? Is the acceleration constant? Are the forces constant? Is momentum conserved? Is energy conserved? Is angular momentum conserved? Is the body in free fall? Is it going in a circle? At what time in its motion do I have information I can use? How many different times are involved?" You are squeezing new knowledge out of the statement of the problem, new knowledge that will provide just the idea necessary to solve the problem.

Classify the problem

You have prepared yourself for choosing equations for problems by classifiing and describing each equation at the time you learned it. Now classify the problem.

Does the problem involve:

> motion with constant acceleration
> motion in a circle
> forces in equilibrium
> forces that change with time
> motion in two dimensions
> motion of a rigid body
> motion under a gravitational force

or one of many other categories?

Compare your classification of the problem with your list of classified equations to eliminate all equations but a few. Of these choose the equations that connect quantities given in the data equations with each other, equations that connect the desired results with each other, or equations that connect the data with the results.

One additional classification of equations that is useful for homework and on tests is to choose equations from the chapter of the book containing the problem. This classification of equations is also useful when you are helping someone else solve a problem that comes from a textbook that you are not using,

Classify the variables

When you write down an equation that is a candidate for your problem, underline the quantities in the equation that are given in the statement of the problem or can be looked up and are therefore known. Circle the quantities that are asked for as one of the results of the problem. Then count the number of quantities that are neither given nor asked for. If the number of extra quantities is zero, you have chosen the right equation, but check again that it holds under the conditions of the problem. If the number of extra quantities is one, but the extra quantity appears in two different equations, you can eliminate it and solve the problem. If you have more extra quantities than equations, look for another equation that holds under the conditions of the problem or reread the problem to make sure that you did not miss a given quantity. (Did you write down $v_0 = 0$ when you read "starts from rest"?)

There does not have to be a single step or a single equation that connects the data directly with the results. A chain of equations connecting quantities each of which depends on the next will do. This is not the place to worry about whether you can do the algebra that will eventually connect the data with the results. That comes later.

General equations

Before you write down a chosen equation, write the reason that you think that equation is appropriate. If the acceleration of an object in the problem is constant, and you are going to use one of the equations that applies to motion under constant acceleration, write "for constant acceleration." If energy is conserved, write "since energy is conserved." Conditions that are not written down cannot be checked for errors and do not help you to think about the problem. The assumption that is not written down will be the one that is wrong.

Write each equation first in the form in which it is famous, or in the form in which you found it. Just by writing an equation in its famous form on a test you get credit for knowing something. In professional work write the reference to the book or paper in which you found each equation, since an esoteric equation whose parentage you can not remember is annoying when you come upon it later:

For constant acceleration:

$\mathbf{a}$ = constant

$\mathbf{v} = \mathbf{v}_0 + \mathbf{a}t$

$\mathbf{r} = \mathbf{r}_0 + \mathbf{v}_0 t + \frac{1}{2}\mathbf{a}t^2$

In this case the general equations are vector equations. Unless you can solve the problem by using vector algebra, a general vector equation needs to be written as three particular equations for the three components of the vectors. The next section shows an example of this.

Particular equations

After writing the conditions under which an equation holds and then the general and famous form of the equation just as it came from the book, write the equation again using the particular variables that you have defined for your problem. Now x becomes x_1 or x_2 or even d, y becomes y_1 or h, and so on.

In the case of vector equations write the separate equations for the components of the vectors.

For constant acceleration:

$v_x \qquad = v_{0x} + a_x t$

$v_y \qquad = v_{0y} + a_y t$

$v_z \qquad = v_{0z} + a_z t$

$x \qquad = x_0 + v_{0x} t + \frac{1}{2} a_x t^2$

$y \qquad = y_0 + v_{0y} t + \frac{1}{2} a_y t^2$

$z \qquad = z_0 + v_{0z} t + \frac{1}{2} a_z t^2$

Algebra

Together with setting up a complete definition and using appropriate equations, doing algebra correctly is an essential part of solving a problem. There are four common troubles when doing algebra: not knowing which operation to use to get where you want to go, using operations incorrectly, inventing operations that are wrong, and copying symbols incorrectly from one step to the next.

Small number of known operations

To avoid two forms of trouble (using operations incorrectly and inventing operations that are wrong) make a list of all of the algebraic operations that you need: multiplying both sides of an equation by something, adding something to both sides of an equation, multiplying a sum of terms by another sum of terms, and so forth. Eliminate the operations that are made up of a series of simpler operations. (One step at a time.) The remaining operations are your algebraic tool kit. Vow to use only these operations for the rest of your life.

Many troubles with algebra come from sudden on-the-spot invention of new algebraic operations. If you suddenly decide, in the heat of solving a problem, that

$$\mathbf{?} \qquad \frac{1}{a+b} = \frac{1}{a} + \frac{1}{b} \qquad \mathbf{?}$$

you have invented an operation that does not exist. Stick to a small number of algebraic operations, do them one at a time, and be sure not to invent new ones, no matter how useful they may seem.

Copying symbols

Copying a symbol incorrectly from one equation to the next can make an entire solution wrong. Special things to watch for are:

Exponents: it is easy to write x instead of x^2,

Subscripts: it is easy to write x_a instead of x_b

Negative signs: $-$ is small and inconspicuous. Treat it with respect. Even the most experienced scientists and engineers have trouble here.

Get unknowns on left hand side alone

The goal is to get each final unknown by itself on the left hand side of an equation with the right hand side containing only quantities whose values are known. This usually has to be done in several steps. First get the unknown on the left hand side by itself:

$$x_1 \qquad = \tfrac{1}{2} a_x t_1{}^2$$

Then use other equations to get rid of intermediate unknowns on the right hand side:

$$x_1 \qquad = \tfrac{1}{2} a_x t_1{}^2$$

$$= \tfrac{1}{2} \, a_x \left(\frac{v_1}{a_x}\right)^2$$

$$= \tfrac{1}{2} \frac{1}{a_x} \, (v_1{}^2)$$

$$= \tfrac{1}{2} \frac{1}{a_x} \left(\frac{W}{K\cos\theta}\right)$$

There are two kinds of unknown quantities. The first kind is one of the results toward which you are working:

To find x:

For motion under constant acceleration:

$$x \qquad = x_0 + v_0 t + \tfrac{1}{2} a t^2$$

$$x_1 \qquad = 0 + 0 + \tfrac{1}{2} a_x t_1{}^2$$

$$= \tfrac{1}{2} a_x t_1{}^2$$

The second kind of unknown is an intermediate result that needs to be found in order to calculate the first kind:

To find t:

For motion under constant acceleration:

$$v \quad = v_0 + at$$

$$v_1 \quad = 0 + a_x t_1$$

$$t_1 \quad = \frac{v_1}{a_x}$$

$$\quad = \frac{6.01\,\mathrm{m/s}}{0.6\,\mathrm{m/s}^2}$$

$$\quad = 10.02 \text{ sec}$$

While you must calculate the values of the final results, you have a choice with the intermediate results. You can calculate their numerical values, or you can substitute their algebraic expressions into the expressions for the final results. Each choice has some advantages. It is worthwhile to calculate intermediate results, find their values, and think about whether they are reasonable. This is one place where numbers can be entered slightly before you get to the end of the problem. Calculating intermediate quantities provides a useful check and also simplifies the equations for the final quantities that you are trying to find. Keeping t, whose value you have calculated, in the equation for x_1 makes that equation less complicated.

On the other hand, putting the algebraic expression for an intermediate result into the final equation can lead to important discoveries. If things cancel each other, the final equation becomes simpler. Sometimes, you can learn that the result does not even depend on one of the things that you thought it did. Simplifying the result in this way is an important part of understanding the problem.

Since calculating the numerical values of intermediate results is useful, and keeping the algebraic expressions for intermediate results is also useful, experienced problem solvers often do both. If their final equations simplify, good. If not, they go back and use the numerical values.

Check units

When you have an equation with an unknown quantity together with its unit alone on the left hand side, put in the units of all of the quantities on the right hand side and check for agreement. If the left hand and right hand units do not agree, the equation is not correct. Stop and correct it before going any farther.

Put in known numbers

After you have written the result algebraically and have checked the unit, put in the known numbers (all of which come from the data equations at the beginning of the solution or from the calculation of intermediate values). To make it clear which symbol each number is replacing, write the numbers in the same relative positions as the symbols they replace. An equation containing numbers alone with no equation above it showing the corresponding symbols is difficult to understand even a few minutes after it is written.

To find x:

For motion under constant acceleration:

$$x \qquad = x_0 + v_0 t + \tfrac{1}{2}at^2$$

$$x_1 \qquad = 0 + 0 + \tfrac{1}{2}a_x t_1^2$$

$$= \tfrac{1}{2}a_x t_1^2$$

$$= \tfrac{1}{2}(0.6\,\text{m/s}^2)(10.02\text{s})^2$$

$$= 30.1 \text{ m}$$

Did you see the squared number? It is easy to miss the exponent and get a wrong answer in the very last step.

Significant figures

Use approximately the same number of significant figures for the result as the number of significant figures used in the original data for the problem. If the time is $t=3.1$ s and the distance is $d=45$ m, the velocity is $v=14.5$ m/s, not 14.516129032 m/s. Since it is easy to calculate with high accuracy, it is wise to use two more significant figures during the intermediate calculations than contained in the original data. If you need to look up data that were not given in the statement of the problem, two additional significant figures keep you safe. If a radius is 45 m and you need to use the value of π, use $\pi=3.142$. The extra figures prevent unnecessary inaccuracy. However, the final result should contain approximately the number of significant figures that the data contained.

Put in units

After the numbers (or with the numbers) come the units:

$$x_1 \quad = \tfrac{1}{2}a_x t_1^2$$

$$= \tfrac{1}{2}(0.6\text{m/s}^2)(10.02\text{s})^2$$

$$= 30.1 \text{ m}$$

A numerical result without a unit means nothing.

Results

Present your results so that they can be identified instantly.

In a five page calculation the results may take up only one line, but the person reading the calculation wants to see that one line first. Make it easy to find. Write the result slightly larger than the rest of the calculation and put a box around it. The result should be an equation with a definition, an equals sign, a symbol, an equals sign, and a number and its unit on the right hand side:

$$\boxed{\text{takeoff distance} = x_1 = 30.1 \text{ meters}}$$

Although the definition is the same one you used before in your list of symbols, it is handy to have it written down again close to the answer. It is easy to forget later exactly what quantity you were calculating.

Rewrite

Rewrite the whole solution. After you have the answer the problem becomes much clearer than when you began to work on it. Now you can write the whole solution in a clear logical way that will show the instructor the best that you can do and will also be understandable later when you are studying for the final exam. If it is a professional problem, this rewrite will be made into a report that is understandable and useful to your supervisor and to yourself at a later time.

One exception to rewriting is on exams. Here rewriting can be confusing to the person grading the paper and should be avoided. Since time is limited on an exam, the time saved by not rewriting can be spent doing the solution neatly and carefully the first time.

Check the solution

Every calculation needs to be checked. If a small calculation has ten steps and the probability of getting each step right is 95%, the probability of getting the whole calculation right is 60%. The best way to check a calculation is to do it again in an entirely different way. If you do not know of a different way to do it, check by redoing it the same way, but realize that it is very difficult to see mistakes once they are made. Everything usually looks all right.

Reread the problem

Go back to the original statement of the problem and check that you have done the problem that is asked for. You have already done this once before, after you put all of the given information down on your paper, but now that you have worked the whole problem and have seen the result, you understand the problem better and should check again.

Check algebra

Check each algebraic step. Do each one a different way if you can. Be especially careful with exponents. With scientific notation a mistake in an exponent makes your answer ten times more wrong than a mistake elsewhere.

Units

One of the best checks for algebraic mistakes is the unit of the result. If original equations are right and the unit is wrong, there is an algebraic mistake.

Variational check

Check the algebraic form of the answer, before the numbers were installed, to see how the left hand side changes when quantities on the right hand side change:

$$x_1 \quad = \frac{1}{2} a_x t_1^2$$

$$= \frac{1}{2} a_x \left(\frac{v_1}{a_x}\right)^2 \qquad = \frac{1}{2} \frac{1}{a_x} (v_1^2)$$

$$= \frac{1}{2} \frac{1}{a_x} \left(\frac{W}{K\cos\theta}\right)$$

Should the runway length x_1 increase as the weight W increases?
Should the length increase as the angle θ between the force and the vertical increases?
Should the length decrease as the force constant K increases?
Should the length decrease as the acceleration down the runway a_x increases?

If the variation does not seem right, there may be an algebraic mistake.

Test cases

Put in test values of the numbers, not the ones in the actual problem, to see if the result makes sense with these test values.

Zero

Zero is one of the best test values. Let each variable go to zero and think about what that should do to the result. Put zero in, and see if what you expected happens.

Infinity

Sometimes infinity is a good test value. Think about what the result should be if one of the variables goes to infinity. Put in infinity for that variable and see if the result does what you expect.

Size of result

In some cases, where you have some experience with the quantities you are calculating, judge whether the size of the result is reasonable. Again, check the exponent of the result more carefully than anything else. 2.99×10^9 m/s is much farther away from the velocity of light than is 2.89×10^8 m/s.

Mistakes

The three reasons for working problems clearly, step by step, are:

> to help in thinking about the problem and working
> toward a solution,
> to be able to understand the problem when you
> come back to it much later, and
> to make it possible to find mistakes.

Mistakes are not completely avoidable. Of all the people who do calculations, only a small number work without making mistakes. The methods we have been discussing will help you to work in a way that minimizes mistakes and that will allow you to find and correct them.

The mistake that is not written down will never be found (except by your instructor or supervisor). By writing every step of a solution, putting every condition in explicitly, and keeping nothing in your head, you at least give yourself a chance of finding the mistakes. A wrong assumption that is not written down is a dangerous thing.

Once a mistake is made, it is not easy to find. Editors and proofreaders know that an author can read his own work several times and still not find simple mistakes in spelling and grammar. Our eyes see the right word even when the wrong word is on the page. Finding our own mistakes is certainly no easier with equations. That is why it is better to redo the solution a different way than to go over it again and again looking for mistakes.

At the beginning of the solution of any problem everyone goes up a few wrong paths. If you make no mistakes of this kind, you are not doing hard enough problems. When you find the right path, solve the problem, recopy it, and hand it in, the wrong paths will not show. Wrong paths are not in the same category as mistakes in calculation that are caused by not following the rules of algebra or by not working in an orderly way.

CHAPTER

6

Can't Solve It

The drawing is beautiful, the data equations are all written down, the geometry equations are solved, and you still don't know how to do the problem. Now what?

Use thinking where it counts

When you are attacking a problem that you do not know how to do and the automatic activities are over, there are few substitutes for thinking. Many of the how-to-solve-it books listed at the end of this book discuss the kind of thinking that is needed. We give just a brief summary here.

Look for a similar problem

Look in the textbook or in the literature for a similar problem that has been solved. Seeing the solution of a similar problem can give you ideas on how to proceed. The library usually has a collection of introductory textbooks, all of which contain solved examples. Looking at the way material is presented in other textbooks can often help you to get through a difficult course. Sometimes, doing the problem in your textbook that comes before or after the one assigned can give you a hint and get you started.

Simplify

Make up a similar problem that is simpler than the problem that you want to solve. Working on a simpler problem can give you a start and can show you the structure of the more complicated problem.

Generalize

Sometimes progress can be made by generalizing the problem. Remove some of the constraints until the problem is very general. This is only useful in some cases, but when it works it lets you solve a whole class of problems instead of one particular problem.

Put in numbers

If you feel that you have found the right equations that connect the known and unknown quantities in a problem, but cannot combine the equations to find the answer, putting in numbers can help. You may be used to solving

$$3x - 5y = 5$$
$$7x + 2y = 3$$

for x and y, but may not be experienced with

$$ax - by = c$$
$$dx + ey = f \ .$$

Although the preferred way of solving a problem is to use symbols for the known and unknown quantities until the last step, you may increase your chances of success with equations by putting in the known numbers. If you have trouble at the beginning of a course, by all means put in numbers. As you become more experienced, try to use symbols instead.

When a problem contains no known quantities at all, using simple numbers, 1, 10, etc., in place of the abstract quantities described in the problem can make the ideas in the problem easier to understand. Going 10 miles in one hour is easier to understand than is $v = d/t$. Once you straighten out the ideas using numbers, you may be able to solve the problem using symbols.

Look for unused data

If you have done part of a problem and cannot do the rest, compare the data given in the statement of the problem with the data you have used. The data that were given but not used may provide the clue to solving the remainder of the problem.

Try a ratio

If you seem not to have enough information to find a quantity that is asked for, try dividing it by a quantity of the same type. That is, divide a mass by a mass, a velocity by a velocity, and so on. Sometimes the same unknown quantity appears in the top and bottom of the right hand side of the ratio equation and cancels.

Put it aside

When you have described a problem on paper and understood it as well as you can, but still cannot solve it, put it aside for a short time, or better, overnight. If you have worked hard enough on a problem and thought about it for a long enough time, sometimes the back of your mind can solve a problem that the front cannot. The larger a problem is, the better this works.

Go for a little help

After drawing a diagram, making a list of definitions and symbols, writing the data equations, converting the units, writing the preliminary equations, looking for a similar problem, simplifying the problem, and generalizing the problem, you still do not know how to solve it, go for a little help.

Find a teacher, a learning center tutor, or a colleague who is willing to act as a consultant. It is not necessary that they know how to do the problem, but it is better if they have experience doing problems.

Asking someone for help with a problem is good practice for what you will be doing as a scientist or engineer. Asking for help does not show weakness. It shows that you are willing to learn, that you realize that you cannot know everything, and that you know the most efficient way to get your work done. Since asking for help requires you to explain your problem to someone else, it forces you to state the problem clearly. In fact, describing a problem in a simple understandable way to someone who knows nothing about the subject can unlock your creativity in finding a solution.

How to ask for help

By going to a teacher for help you will learn how to present an unfinished problem to a consultant in an efficient way. You have only a few seconds in which to introduce yourself by name, state the problem, show your preparations, and ask for advice. As soon as you get a new idea, leave, and try to finish the problem by yourself. Going back for help a second time shows commitment and tenacity.

Introduce yourself by name. Saying, "Hello, I'm Alexander Alexandervich," is not only polite, but also avoids the awkwardness some teachers feel when having to ask you your name. Rather than saying, "I can't do problem seven," show the original statement of the problem, show your drawing, definition list, and other preparations, and discuss your ideas.

As soon as your consultant provides a single new idea that you think may help, leave. Go away and work on the problem again yourself. If there is still trouble, go again for a little help. As soon as a single useful idea surfaces, go away and work by yourself.

If you allow someone else to solve your problem, that problem is spoiled forever. You will never understand every step until you have invented and thought about every step yourself. Those steps that someone else does are just the ones you will not understand on the next problem or on the exam. Get help, but do the problem yourself.

Going to a teacher for help is always a good idea even if the teacher is busy or is difficult to understand. Most teachers will be glad to see you a few times a semester. Even if you display ignorance about the material in the course, you will be displaying interest in it. Sometimes teachers are unaware of the difficulties students are having with assigned problems. Asking a teacher a question about a problem can give her useful information and help everyone in the course.

CHAPTER

7

□□□□□□ **7** □□□□□

Spreadsheets

Spreadsheets, originally developed for financial calculations, have been adopted by scientists and engineers to do calculations, to organize experimental data, and to make graphs.

A spreadsheet is an array of numbered rows (across) and lettered columns (down) of cells, each of which can hold a number, an equation, a name, or a comment. A spreadsheet is an orderly way to set up a calculation. It encourages breaking a problem into parts and writing the parts separately. A spreadsheet can be made self documenting, displaying separately the definition, name, value, and unit of each variable. Each step in the calculation is displayed and can be commented on, making the calculation easy to understand and errors easy to find.

When an equation needs to be calculated for many values of one of its variables in order to plot a graph or do an integral numerically, an array is the best way to display the results. The values of the independent variable march down the page, while the building up of the calculation necessary to get the results spreads across the page. The values of the independent variable are clearly displayed in the first column, and each successive column can be checked for unexpected zeros or infinities.

Spreadsheets are a natural way to set up matrix calculations, especially those in which successive matrices act on a column vector, such as in geometrical optics and polarization calculations.

Computers have made spreadsheets into powerful calculational tools. The computer can calculate the left hand side of each equation on the spreadsheet as long as the value of every quantity on the right hand side exists somewhere on the spreadsheet. A computer spreadsheet not only helps to organize and document the solution to a problem but does all of the arithmetic correctly as well.

Single values

For calculations that have only one value for each variable and therefore are not repetitive but require many steps to get the answer, the successive steps can spread down the page, with each step on a separate line. Each line contains the definition of a quantity, its name, its value or equation, and its unit.

The next two sections are an example of a single-value problem and its solution worked out in spreadsheet form.

Example 3: Solar rocks

A solar collector absorbs 18 kilowatts of solar power for a period of 6 hours. The collected energy is used to heat a metric ton (1000 kg) of rocks whose specific heat is 0.5. How much does the collected solar energy increase the temperature of the rocks?

Solution for example 3

Here are the equations for the solution:

	A	B	C	D
1	**Solar Rocks**			
2	D. Scarl			
3	31047			
4	**Definitions and Data:**			
5	power absorbed	Pkw	18	kw
6	time period	th	6	hours
7	mass of rocks	mt	1	ton
8	specific heat of rocks	s	0.5	

9	Constants:			
10	specific heat capacity of water	cw	4190	joules/(kg-K)
11	**Unit Conversion:**			
12	power absorbed	PJ	=Pkw*1000	watts
13	time period	ts	=th*3600	sec
14	mass of rocks	mkg	=mt*1000	kg
15	**Preliminary Equations:**			
16	energy absorbed	Q	=PJ*ts	joules
17	specific heat capacity of rocks	cr	=s*cw	joules/(kg-K)
18	**Result:**			
19	temperature increase of rocks	Tr	=Q/(cr*mkg)	K

Here are the results of the equations as calculated by a computer spreadsheet:

	A	B	C	D
1	**Solar Rocks**			
2	D. Scarl			
3	Jan 1, 1989			
4	**Definitions and Data:**			
5	power absorbed	Pkw	18	kw
6	time period	th	6	hours
7	mass of rocks	mt	1	ton
8	specific heat of rocks	s	0.50	
9	**Constants:**			
10	specific heat capacity of water	cw	4190	joules/(kg-K)
11	**Unit Conversion:**			
12	power absorbed	PJ	18000	watts
13	time period	ts	21600	sec
14	mass of rocks	mkg	1000	kg
15	**Preliminary Equations:**			
16	energy absorbed	Q	3.89E+08	joules
17	specific heat capacity of rocks	cr	2095	joules/(kg-K)
18	**Result:**			
19	temperature increase of rocks	Tr	185.6	K

Inputs

The first section of the spreadsheet contains all of the inputs. These are the data equations of the problem. Each input equation is on a separate line with the definition in column A, the name in column B, the value in column C, and the unit in column D:

	A	B	C	D
4	**Definitions and Data:**			
5	power absorbed	Pkw	18	kw
6	time period	th	6	hours
7	mass of rocks	mt	1	ton
8	specific heat of rocks	s	0.5	

Constants

Constants are quantities that are not given in the statement of the problem, but are needed for its solution. They can be properties of materials, like density, or fundamental constants, like Planck's constant. Each is on a separate line that includes the description of the constant, its name, its value, and its unit:

	A	B	C	D
9	**Constants:**			
10	specific heat capacity of water	cw	4190	joules/(kg-K)

Unit conversion

If the units given in the statement of the problem are not the units needed to solve the problem, this set of spreadsheet rows converts the given units into standard SI units.

	A	B	C	D
11	**Unit Conversion:**			
12	power absorbed	PJ	=Pkw*1000	watts
13	time period	ts	=th*3600	sec
14	mass of rocks	mkg	=mt*1000	kg

When the unit of a quantity changes on a spreadsheet, its name must change too. The spreadsheet, like a computer program, can only accept one value for each name.

Calculation

The steps of the calculation can be done row by row down the spreadsheet so that the definitions, the data equations, and the calculation all have almost the same format:

	A	B	C	D
15	**Preliminary Equations:**			
16	energy absorbed	Q	=PJ*ts	joules
17	specific heat capacity of rocks	cr	=s*cw	joules/(kg-K)

Results

The last few rows of the spreadsheet present the results. Again, each result has its definition in the first column, its name in the second, its value in the third, and its unit in the fourth.

Typed as:

	A	B	C	D
18	**Result:**			
19	temperature increase of rocks	Tr	=Q/(cr*mkg)	K

Displayed as:

	A	B	C	D
18	**Result:**			
19	temperature increase of rocks	Tr	185.6	K

Functions

When the dependent variable has to be calculated for many different values of the independent variable in order to calculate a function such as $y(x)$ or to make a graph, the calculation is best done across the spreadsheet. The first column contains all of the values of the independent variable x and succeeding columns are used to build up the calculation of the dependent variable y.

The next two sections are an example of a problem and its spreadsheet solution, set up, calculated, and graphed.

Example 4: Distance vs. angle for a thrown baseball

An outfielder can throw a baseball with a speed of 70 miles an hour. Make a graph showing how the distance he can throw the ball depends on the angle at which he throws it.

Solution for example 4

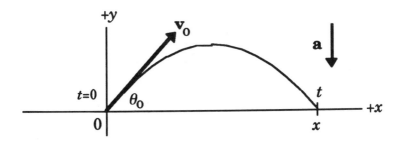

The first section of the spreadsheet is again a list of definitions of symbols:

	A	B	C	D
1	**Maximum Throwing**			
2	**Distance**			
3	D. Scarl			
4	31048			
5	**Definitions and data:**			
6	angle above horizontal	thetad		degrees
7	time of landing	t		s
8	starting distance	xzero	0	meters
9	distance thrown	x		meters
10	initial height	yzero	0	meters
11	final height	y	0	meters
12	initial velocity	vmph	70	mi/hr
13	horizontal initial velocity	v0x		m/s
14	vertical initial velocity	v0y		m/s

15	horizontal acceleration	ax	0	m/s^2
16	vertical acceleration	ay	=–g	m/s^2
17	**Constants:**			
18	acceleration of gravity	g	9.807	m/s^2
19	**Unit conversion:**			
20	initial velocity	vmps	=vmph*1610/3600	m/s
21	angle above horizontal	theta		radians

The next section is a column listing the throwing angles, from 0 (horizontal) to 90 degrees (straight up), followed by columns calculating the variables that depend on these angles:

	A	B	C	D	E	F
22	**Calculations**					
23	thetad	theta=	v0x=	v0y=	t=	x=
24		thetad	vmps	vmps	-2*vy/ay	v0x*t
25		*pi()/180	*cos(thetad)	*sin(thetad)		
26						
28	0	=A28*PI()/180	=vmps*COS(B28)	=vmps*SIN(B28)	=-2*D28/ay	=C28*E28
29	5	=A29*PI()/180	=vmps*COS(B29)	=vmps*SIN(B29)	=-2*D29/ay	=C29*E29
30	10	=A30*PI()/180	=vmps*COS(B30)	=vmps*SIN(B30)	=-2*D30/ay	=C30*E30
31	15	=A31*PI()/180	=vmps*COS(B31)	=vmps*SIN(B31)	=-2*D31/ay	=C31*E31
32	20	=A32*PI()/180	=vmps*COS(B32)	=vmps*SIN(B32)	=-2*D32/ay	=C32*E32
33	25	=A33*PI()/180	=vmps*COS(B33)	=vmps*SIN(B33)	=-2*D33/ay	=C33*E33
34	30	=A34*PI()/180	=vmps*COS(B34)	=vmps*SIN(B34)	=-2*D34/ay	=C34*E34
35	35	=A35*PI()/180	=vmps*COS(B35)	=vmps*SIN(B35)	=-2*D35/ay	=C35*E35
36	40	=A36*PI()/180	=vmps*COS(B36)	=vmps*SIN(B36)	=-2*D36/ay	=C36*E36
37	45	=A37*PI()/180	=vmps*COS(B37)	=vmps*SIN(B37)	=-2*D37/ay	=C37*E37
38	50	=A38*PI()/180	=vmps*COS(B38)	=vmps*SIN(B38)	=-2*D38/ay	=C38*E38
39	55	=A39*PI()/180	=vmps*COS(B39)	=vmps*SIN(B39)	=-2*D39/ay	=C39*E39
40	60	=A40*PI()/180	=vmps*COS(B40)	=vmps*SIN(B40)	=-2*D40/ay	=C40*E40
41	65	=A41*PI()/180	=vmps*COS(B41)	=vmps*SIN(B41)	=-2*D41/ay	=C41*E41
42	70	=A42*PI()/180	=vmps*COS(B42)	=vmps*SIN(B42)	=-2*D42/ay	=C42*E42
43	75	=A43*PI()/180	=vmps*COS(B43)	=vmps*SIN(B43)	=-2*D43/ay	=C43*E43
44	80	=A44*PI()/180	=vmps*COS(B44)	=vmps*SIN(B44)	=-2*D44/ay	=C44*E44
45	85	=A45*PI()/180	=vmps*COS(B45)	=vmps*SIN(B45)	=-2*D45/ay	=C45*E45
46	90	=A46*PI()/180	=vmps*COS(B46)	=vmps*SIN(B46)	=-2*D46/ay	=C46*E46

The first column contains all of the values of the independent variable that you will need. The next columns break the equation for the dependent variable up into small steps. In this example, column A is the angle in degrees, column B is the angle in radians, columns C and D are the horizontal and vertical components of the initial velocity, column E is the time at which the ball lands, and column F is the distance thrown. In each column, all of the equations are identical, except that they refer to different values of the variable in the first column. By calculating like this, all of the partial results are visible for each value of the independent variable. If something goes wrong, or a mistake is made, you can see it right away.

You can also see how the partial results change as the independent variable changes. In this example you can find not only which angle leads to the greatest distance but also which angle keeps the ball in the air for the longest time.

Each column heading contains the name of the variable calculated in that column and a facsimile of the equation used in the rest of the column to calculate that variable. This facsimile equation is part of the documentation of the calculation and is not actually used by the spreadsheet.

On the next page is what the spreadsheet looks like when the formulas are hidden and the calculations are shown. The last column contains the results, which the spreadsheet has plotted against the first column to give a graph of the distance thrown versus the throwing angle..

	A	B	C	D	E	F
22	**Calculations:**					
23	thetad	theta=	v0x=	v0y=	t=	x=
24		thetad	vmps	vmps	-2*v0y	v0x*t
25		*pi()/180	*cos(thetad)	*sin(thetad)	/ay	
26						
28	0	0.00	31.31	0.00	0.00	0.00
29	5	0.09	31.19	2.73	0.56	17.35
30	10	0.17	30.83	5.44	1.11	34.18
31	15	0.26	30.24	8.10	1.65	49.97
32	20	0.35	29.42	10.71	2.18	64.24
33	25	0.44	28.37	13.23	2.70	76.55
34	30	0.52	27.11	15.65	3.19	86.54
35	35	0.61	25.64	17.96	3.66	93.91
36	40	0.70	23.98	20.12	4.10	98.41
37	45	0.79	22.14	22.14	4.51	99.93
38	50	0.87	20.12	23.98	4.89	98.41
39	55	0.96	17.96	25.64	5.23	93.91
40	60	1.05	15.65	27.11	5.53	86.54
41	65	1.13	13.23	28.37	5.79	76.55
42	70	1.22	10.71	29.42	6.00	64.24
43	75	1.31	8.10	30.24	6.17	49.97
44	80	1.40	5.44	30.83	6.29	34.18
45	85	1.48	2.73	31.19	6.36	17.35
46	90	1.57	0.00	31.31	6.38	0.00

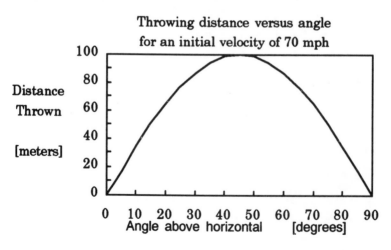

Throwing distance versus angle
for an initial velocity of 70 mph

Integrals

As well as single values and functions, a spreadsheet is useful for calculating numerically the integral of functions that cannot be found in the standard integration tables. Efficient ways to do this, described in books on numerical methods, all require calculation of the value of the function under the integral for many values of the variable over which you want to integrate. These values are then weighted and added. On a spreadsheet the first column contains the values of the integration variable, the next columns break up the function under the integral into simple parts, and the last column contains the weighted values needed by the numerical integration method. At the bottom of the last column a single cell can be used to sum all of the cells in the column and give the value of the integral.

Matrices

Spreadsheets also excel in doing matrix calculations. Matrix calculations require the multiplication of matrices times vectors, of matrices times matrices, or of vectors times vectors. They also include inverting matrices and finding traces, determinants, and eigenvalues. A spreadsheet, which is already an array, is a convenient and obvious way to set up matrix calculations.

Computer spreadsheets

Each cell of a computer spreadsheet can hold either a word or sentence, a number, or an equation. Each equation can use values gotten from other cells. When a cell is changed, the computer automatically recalculates all of the cells and displays the new values. Changing the values in an equation and immediately seeing the results of the change is one of the pleasures of working with computer spreadsheets.

In pen and paper calculations it is best to do the solution algebraically and to insert the data numbers in only the last equations. In a computer spreadsheet calculation, you can see both the algebraic equations and the numerical value of those equations all the way through the calculation. This helps in thinking about the problem and in avoiding mistakes.

When comparing an equation with experimental results, the spreadsheet makes it easy to change theoretical assumptions until the calculated values correspond to the measured values. This is a powerful tool in experimental work.

Since computer spreadsheets have all of the functions that computer languages have, such as sin, exp, rnd, abs, and many others, any equation that can be written in a computer language can be written in a spreadsheet cell. For many calculations computer spreadsheets are easier to use than computer programs. Much of a well structured program consists of formatted input statements, formatted output statements, and DO loops. A computer spreadsheet does all three of these things automatically, leaving you free to work on the structure of the problem and the equations.

A spreadsheet is much easier to debug than a computer program, since the spreadsheet displays all intermediate results, almost like a program with a PRINT statement after each line of the calculation. If a value goes negative or goes to infinity on a spreadsheet, it is immediately obvious, and the cause can be found quickly. A spreadsheet can be debugged in a small fraction of the time it would take to debug the equivalent program.

As an additional bonus, most computer spreadsheets can make a graph by plotting the values in one column against the values in another column. If changes are made in either column, the graph shows those changes immediately.

Spreadsheet equations

An equation typed into a spreadsheet cell appears two ways. While it is being typed and until you press ⌐return⌐, the equation itself appears. As soon as you press ⌐return⌐, the computer calculates the value of the equation and displays the value in the same cell. However, the equation is still there and can be seen by selecting that cell again. The spreadsheet assumes that any cell entry that begins with an arithmetic operator (usually = or +) is an equation.

An equation can use the values in other cells of the spreadsheet by using the locations or the names of those cells. Thus =A1*A2 means multiply the value in cell A1 by the value in cell A2 . If you later change the value in A1, the result of this calculation will change correspondingly. Even more elegantly, cells can be given names, and those names can be used in calculations. If cell A1 is named *velocity* and cell A2 is named *time,* then a cell with the equation *=velocity*time* will give the distance travelled.

Spreadsheet editing

Computer spreadsheets know that you often need to write almost the same equation in many cells. For instance, if column A of the spreadsheet contains values of x from 0 to 1 in steps of 0.1, and you want to plot $1/x^2$, the first cell of column B contains =1/(A1*A1), the second cell contains =1/(A2 *A2), the third cell, =1/(A3*A3), and so forth. All you need to write is the equation in the first cell, =1/(A1*A1). When you copy this equation into the next cell down, the computer automatically moves the reference one cell down so that now the equation reads =1/(A2 *A2). With a single command, you can fill an entire column with the same equation, each row referring to the proper value in the other columns.

The first column, containing equally spaced values of the independent variable, can be generated automatically by the spreadsheet. After being given a starting value and a step value, the spreadsheet will automatically fill a selected column or row with equally spaced numbers.

Organizing a computer spreadsheet

The rules for setting up a spreadsheet to help solve a problem are almost the same as those discussed in the previous chapters for setting up the problem itself. Break the spreadsheet up into parts: first the data given in the original statement of the problem, then constants and other outside values, the preliminary equations, the science equations, and finally, the results.

If an algebraic expression is complicated, calculate each part separately and then combine the parts to get the result. This makes the equation in each cell easy to write, understand, and debug.

Avoid numbers

A good rule for spreadsheet equations (and for equations in any computer program) is to avoid numbers. All of the numbers given in the data and all of the constants should be entered in the first rows of the spreadsheet. These numbers should be referred to by row and column or by name in the equations that follow. If you put a number into an equation in a cell, it usually gets copied into many other cells and is very hard to find and change later. If you put a named variable or a cell reference into an equation, it automatically will change everywhere when you change the value in the one original cell. (Refer to a data or constant cell using an absolute reference that does not change as it is copied from cell to cell.)

Naming variables

Naming variables in a computer spreadsheet is slightly different from naming then in a calculation on paper. Some spreadsheets do not allow subscripts or superscripts, but variables such as v_{x0} can be written as vx0. However, variables made up of one letter and one number are usually not allowed. Since spreadsheet rows are designated by numbers and columns by letters, the name of a cell is a letter followed by a number, and variables are not allowed to have the same names as cells. It is, of course, easy to give variables longer names that also help in recognizing them. Another less satisfactory way around this problem is to ask the spreadsheet to name its cells in the form R1C1, for Row 1 Column 1, etc. This is usually a little awkward, and takes up more room, but R2C5, while it may look like the name of a robot, at least does not look like any normally named physical variable.

CHAPTER

8

Presentation of Results: Reports and Papers

Presenting your results can be as simple as drawing a box around your answer or as complicated as writing a scientific paper. The goal of presenting your results is to tell someone else, or yourself a year later, what problem you have solved and what the answer was. The results should be presented clearly and completely enough that the person reading them does not have to work to figure out what they mean. On homework and tests the results can be an equation with a definition and symbol on the left hand side, a number or algebraic expression on the right hand side, a unit, and a box or other way of attracting the instructor's attention. On the other hand, presenting your results in a report to your supervisor or in a scientific paper requires explaining the problem to someone who does not know as much about it as you do. Brilliant solutions badly presented are often ignored. You will be evaluated in your scientific or technical career not only on your solutions, but also on your presentations.

Title

The title of a report or scientific paper will be read by many more people than any other part. It will be read by people trying to find out if the paper is of enough interest or use to them that they should read farther. The title should say exactly what problem you solved and what about it is different from other similar problems.

Abstract

The abstract will be read by many more people than the paper itself. It will be read by people who were attracted by the title and would like to learn if the paper really is useful to them. The abstract can contain particular details and major results. Since the abstract will be available in abstract journals and computer-searchable bibliographies without the paper itself, it needs to be as self-contained as possible.

Graphs

If the results are more complicated than a single number they are presented in tables or graphs. Graphs summarize a lot of information. They help in visualizing results in a way that lets the reader understand those results and generate new ideas from them. Computers have made graphing easy and quick, but it is still up to you to design understandable graphs.

Axis labels

The x (horizontal) axis is usually used for the independent variable. (In some business oriented spreadsheets, this is called the category axis because it contains the various categories, each of which has a value to be represented on the graph.) The y (vertical) axis is the dependent variable, and one reads a value of y for a given value of x. (In those same spreadsheets, this is called the value axis.)

A label below the x axis tells what variable is actually plotted along the axis, and its unit. A label next to the y axis tells what variable is plotted along that axis and its unit. A graph without these labels is almost impossible to interpret.

Scale divisions

Both the x and y axes need major numbered scale divisions or tic marks and minor unnumbered tic marks. These marks and numbers have to be chosen so that the value of a point on the axis between two of the marked points can be found. Usually the distance between the tics or numbers is a power of 10 times 1, 2, or 5. That is, major divisions can be placed at 0.1, 0.2, 0.3, 0.4...with minor divisions spaced by 0.01 or 0.02 or 0.05. Major divisions can be placed at 10, 20, 40, 60... with minor divisions spaced by 1, 2, or 5, or at 50, 100, 150, 200...with minor divisions spaced by 10, and so forth. It is best to avoid divisions spaced by powers of ten times 2.5 or 4.

If a graph gives detailed information that readers may want to read accurately, put scale divisions on the top and right hand borders of the graph as well as along the axes. The reader can then lay a ruler between the top and bottom divisions and read accurate values from the graph.

Points and lines

The results of a calculation are usually plotted as a continuous line with no symbol marking the individual points. Although the value of y is calculated only at some values of x, the calculated points are joined by a straight line to form the curve. Experimentally measured points are usually plotted as individual points using symbols such as ■ and ▲ .

Straight lines

The easiest graph to understand is a straight line. Sometimes, if the graph is not a straight line, it can be replotted, using different variables, and come out a straight line. For instance $y = x^2$ is not a straight line when y is plotted against x, but becomes a straight line if y is plotted against x^2.

Logarithmic axes

If either x or y covers a wide range (many powers of 10) a graph that shows all of the values will give very little detail about the small values. More information can be gotten from the graph if the axis of the wide-range variable is made logarithmic. This means that instead of its position along the axis being proportional to a value itself, its position is proportional to the logarithm to the base 10 of the value. If 1 inch corresponds to 1, 2 inches corresponds to 100, 3 inches corresponds to 1000, etc. On a linear graph that goes up to 1000, the data between 1 and 10 are hard to see, but on a log graph the data between 1 and 10 are given just as much room as the data between 100 and 1000.

On a log graph, although the distance between 1 and 10 is the same as the distance between 10 and 100, the distance between 1 and 2 is not the same as the distance between 2 and 3. Log graphs have evenly spaced tic marks at .001, .01, 0.1, 1, 10, 100, 1000, etc. and unevenly spaced marks at 0.002, 0.003, 0.004, etc.

On a graph with a logarithmic y axis, the curve $y = e^x$ is a straight line, so that pure exponential functions are easy to recognize on logarithmic graphs.

Problems

After solving each of the first few problems, reread Chapter 3, Chapter 4, and Chapter 5 to compare the style of problem solving that you have used to the style suggested there.

1. Distant lake. An airliner flying at a height of 10,000 m is directly above a small lake. From where you are standing, at the same height as the lake, the angle from the horizon to the airliner is 42°.
a. Write a heading, draw a diagram, and label it with all of the symbols for all of the quantities you will need in order to answer part d.
b. Make a list of definitions, symbols, values, and units for all of the quantities you will need to answer part d.
c. Make the preliminary unit conversion and geometric calculations that will help to solve part d.
d. How far away is the lake?

2. Shortcut. Your college quadrangle is 85 meters long and 66 meters wide. When you are late for class you can walk at 7 miles per hour. You are at one corner of the quad and your class is at the diagonally opposite corner.
a. Write a heading, draw a diagram, and label it with all of the symbols for all of the quantities you will need in order to answer part d.
b. Make a list of definitions, symbols, values, and units for all of the quantities you will need to answer part d.
c. Make the preliminary unit conversion and geometric calculations that will help to solve part d.
d. How much time can you save by cutting across the quad rather than walking around the edge?

3. Satellite period. A satellite in a circular orbit 640 km above the surface of the earth travels at 7.54 km/s.

a. Write a heading, draw a diagram, and label it with all of the symbols for all of the quantities you will need in order to answer part d.

b. Make a list of definitions, symbols, values, and units for all of the quantities you will need to answer part d.

c. Make the preliminary unit conversion and geometric calculations that will help to solve part d.

d. How long does it take the satellite to make one trip around the earth?

4. Building heat loss. A flat-roofed apartment building is 20 m wide, 23 m deep and 27 m high. On a winter day it loses heat through its surfaces exposed to the air at an average rate of 10 joules per m^2 per second.

a. Write a heading, draw a diagram of the building, and label it with symbols for all of the quantities you will need in order to answer part d.

b. Make a list of definitions, symbols, values, and units for all of the quantities you will need to answer part d.

c. Write the preliminary geometric equations that will help to answer part d.

d. How much energy (in joules) does the building lose each day?

5. Ping-pong balls on swimming pool. You want to keep your swimming pool warm by covering it with a single layer of ping-pong balls, packed as closely as possible. The pool is 5 m by 15 m and each ball is 2.5 cm in diameter.

a. Write a heading, draw a diagram, and label it with all of the symbols for all of the quantities you will need in order to answer part d.

b. Make a list of definitions, symbols, values, and units for all of the quantities you will need to answer part d.

c. Make the preliminary unit conversion and geometric calculations that will help to solve part d.

d. How many balls cover the pool and what fraction of the area of the pool is covered by the balls?

6. Filling stadium. A football stadium with 35,000 seats has three gates. When gate A alone is open it takes 40 minutes to fill the stadium. When gate B alone is open it takes 40 minutes to fill the stadium. When gate C alone is open it takes 60 minutes to fill the stadium. How long does it take to fill the stadium with all three gates open?

7. Memory refresh. A computer memory chip containing one million bits of random access memory must refresh every bit every 5 milliseconds The chip is able to refresh one thousand bits in 50 nanoseconds. It is not available to the computer while it is refreshing itself. What fraction of the time is it available to the computer?

8. Weather balloon. A weather balloon made of rubber with a density of 1100 kg/m^3 has a mass of 14 kg. When inflated with helium its diameter is 8 meters.
a. What is the surface area of the inflated balloon?
b. What is the volume of the rubber of the inflated balloon?
c. What is the thickness of the rubber of the inflated balloon?

9. Satellite height. A satellite in a circular orbit that passes over New York and Los Angeles is on the horizon when seen from New York and when seen from Los Angeles at the same time. What is the height of its orbit above the surface of the earth?

10. Fast car. A new Ferrari weighs 3500 lb and can go from 0 to 60 miles per hour with constant acceleration in 4.2 seconds.
a. What is its acceleration in m/s^2?
b. What is the ratio of a to g, the earth's gravitational constant?
c. What is the forward force on the car while it is accelerating?
d. What is the ratio of the forward force on the car to its weight?
e. With the same acceleration, how long would it take the car to go one-quarter of a mile from a standing start?

11. Travelling piston. A piston in a 2 liter automobile engine travels 60 mm down and 60 mm back up each time the engine crankshaft makes one revolution. When the automobile is travelling at 55 miles per hour, the engine is turning at 2600 revolutions per minute. What is the total distance the piston travels (relative to the cylinder wall) when the car travels 1 meter?

12. Automobile engine efficiency A car engine that can produce a maximum of 120 horsepower burns gasoline and turns the heat from the gasoline into useful work with an efficiency of 11 percent. The gasoline weighs 0.9 kg per liter and has a heat energy content of 4.5×10^7 joules per kilogram. How long can the engine run on one liter of gasoline while producing its maximum horsepower?

13. Pulsed laser. A laser turns electrical energy into light with an efficiency of 1%. A 1000 microfarad capacitor charged to 5000 volts transfers its stored electrical energy of 1.25×10^4 joules into the laser, which then produces a uniform 2×10^{-8} second long light pulse.
a. What is the energy in the resulting light pulse?
b. What is the light power during the pulse?

14. Tire wear. An automobile tire wears off 1/2 inch of tread thickness while travelling 50,000 miles. What is the average thickness (in meters) worn off in each revolution of the tire?

15. Passing cars. A Porsche travelling at a constant speed of 90 km/hr comes up behind a Corvette stopped at a light. When the Porsche is 30 m behind the Corvette, the light changes and the Corvette accelerates at 0.8 g. (g is the earth's gravitational constant.)
a. How far from the light does the Porsche pass the Corvette?
b. How far from the light does the Corvette pass the Porsche?

16. First ball. The mayor of Los Angeles throws out the first ball of the season from his box seat 55 meters (horizontally) from the pitchers mound and 8 m above the level of the mound. He throws the ball at an angle of 20° above horizontal. At what speed should he throw the ball so that it will hit the ground at the mound?

17. Human-powered plane. A 220 lb human-powered airplane accelerates from rest along its runway with a constant acceleration of 0.5 m/s^2. The force on its wings is Kv^2 where v is the velocity of the plane and K=30 N-s^2/m^2. The plane just leaves the ground 33 meters from where it started. In what direction is the force on the wings?

18. Orbit radius and period. A satellite with an orbit radius of 10000 km makes one complete trip around the earth in 9940 seconds. How long does it take a satellite with an orbit radius of 20000 km to make one complete trip? (It is not necessary to look up the mass of the earth or the gravitational constant to do this problem.)

19. Earthly speed. The earth revolves on its axis once each day while it circles (ellipses?) the sun once each year. Seen from the north pole of the earth, both of these motions are counterclockwise.
a. How fast does the center of the earth move in its orbit about the sun?
b. With respect to the center of the earth how fast does a person standing on the equator move because of the earth's rotation?
c. How fast does a person at your latitude move?
d. Assuming that the earth's axis is perpendicular to its orbit plane, what are your maximum and minimum speeds with respect to the sun?
e. At what time of day do you have your maximum speed?

20. Pole vault height. The height a pole vaulter can clear is mostly determined by the speed of his initial run. If all of his running kinetic energy is converted into potential energy during the jump, plot a graph of the height (in meters) he can clear vs his horizontal velocity (in meters/second) when he leaves the ground. Use a reasonable range for his velocity.

Further Reading

Some books that are close in level and emphasis to "How to Solve Problems" are:

Becoming a Master Student, 5th Edition, David B. Ellis, College Survival, Inc., P. O. Box 8306, Rapid City, SD 57709, 1985. Although many believe that being a student, writing English, and driving a car are skills we all are born with, they are not. If you do not have this first-rate book, get it, and learn how to be a student .

Used Math; For the First Two Years of College Science, Clifford E. Swartz, Prentice Hall, 1973. A useful handbook of introductory applied mathematics containing the tools one needs to solve beginning science and engineering problems with confidence.

Succeed With Math; Every Student's Guide to Conquering Math Anxiety, Sheila Tobias, College Entrance Examination Board, P. O. Box 886, New York NY 10101-0886, 1987. An even more useful handbook of introductory applied mathematics written with sensitivity and style. Worth owning.

Overview Case Study; Physics Study Guide, Alan Van Heuvelen, Department of Physics, New Mexico State University, Las Cruces, NM 88003. 1990. A workbook that includes templates for getting started in problem solving and much good advice.

The definition of problem solving is expanded to the creative solution of mathematical problems by:

The Art of Problem Posing, Stephen I. Brown and Marion I.
  Walter, Lawrence Erlbaum Associates, 365 Broadway,
  Hillsdale, NJ 07642, 1983. Describes how good
  mathematicians learn to see problems where no one knows
  that there are problems, by deriving general rules from
  particular examples. Shows that a problem is well solved if
  the solution leads to an even more important unsolved
  problem. Essential ideas for those who plan to become
  mathematicians.

How to Solve It, 2nd edition, George Polya, Princeton University
  Press, 1973. The granddaddy of them all. This discussion of
  mostly geometric and algebraic problem solving has led to
  most of the other books on problem solving. A how to do it
  book that contains every method one can think of.

Mathematical Discovery, George Polya, John Wiley and Sons,
  1981. Greatly expanded how-to-solve-it book aimed at
  teachers of mathematics and of problem solving. Many
  mathematical examples.

Mathematical Problem Solving, Alan H. Schoenfeld, Academic
  Press, 1985. Scholarly advanced treatise with a scholarly
  advanced style. Good blow-by-blow descriptions of students'
  attempts to solve geometrical problems with an analysis of
  the thought processes of different types of problem solvers.

Problem Solving and Comprehension, Arthur Whimbey and Jack
  Lochhead, Lawrence Erlbaum Associates, 365 Broadway,
  Hillsdale, NJ 07642, 1986. Shows methods to use on math
  and logic problems of the type found on the SAT's, GRE's,
  and other tests. Can only help.

More general advice for success in research and engineering can
be found in:

The Art of Scientific Investigation, 3rd edition, W. I. B.
  Beveridge, Random House, 1957. Beautifully written
  introduction to scientific thinking, containing many
  descriptions of how well-known scientists worked toward
  their discoveries. The best place to learn the real rewards of
  doing science.

Being Successful as an Engineer, W. H. Roadstrum, Engineering
Press, P. O. Box 5, San Jose, CA 95103, 1978. A good
introduction for people beginning an engineering career.
Covers everything from proposals through research,
manufacturing, and quality control with emphasis on the
skillful management of an engineering group.

An Introduction to Scientific Research, E. Bright Wilson, Jr.,
McGraw Hill, 1952. A wonderful handbook for Ph.D. students
in the sciences. Goes from the choice of a research problem,
through the design of experiments, statistics and data
reduction, mathematical calculations, to the reporting of
scientific results.

James Adams's books are an introduction to the large
literature on creative thinking in engineering, science, art, music,
and everyday life:

Conceptual Blockbusting, 2nd edition, James L. Adams, W. W.
Norton & Co., 1979. A book, written by an engineer, that is a
best seller among managers, Adams encourages solving
problems by taking creative and imaginative end runs
around them. Look at this book to see that "solving
problems" has many meanings, only one of which is
discussed in "How to Solve Problems."

**The Care and Feeding of Ideas; A Guide to Encouraging
Creativity,** James L. Adams, Addison Wesley, 1986. Very
broad non-technical discussions of how to solve problems that
have not yet been formulated. Emphasizes freeing the mind
from ruts and nurturing partially formed thoughts.

After you have solved a problem and want to describe it to
others in clear English read:

The Elements of Style, 2nd edition, William Strunk, Jr, and E. B.
White, Macmillan Publishing Co., 1972. Buy it. Read it. Use
it.

Problem Solutions

1. Distant Lake
D. Scarl
May 15, 1990
A. DRAWING

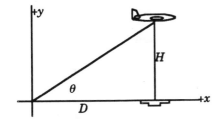

B. DEFINITIONS

Height of airliner	$= H$	$=10\ 000$	m
Angle between horizontal and airliner	$= \theta$	$=42$	°
Distance to lake	$=D$		m

C. GEOMETRY

$$H/D = \sin \theta$$

$$D = \frac{H}{\sin \theta}$$

$$= \frac{10000 \text{ m}}{\sin 42°}$$

$$= \frac{10000 \text{ m}}{0.669}$$

D. $\boxed{\text{Distance to lake} = D = 1.49 \times 10^4 \text{ m}}$

3. Satellite Period
(Name and Date)

A. DRAWING

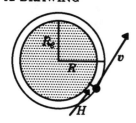

B. DEFINITIONS

Height of satellite above earth's surface $= H$ $=640$ km

Velocity of satellite $= v$ $=7.54$ km/s

Radius of earth $= R_e$ m

Radius of satellite orbit $= R$ m

Period of satellite orbit $= P$ s

DATA

R_e $= 6371$ km   Handbook of Chemistry and Physics, 57th Ed. Chemical Rubber Co. 1971. pg F148

C. GEOMETRY

Find the orbit radius

$R = R_e + H$ $= 6371$ km $+ 640$ km $= 7011$ km

Find the period

$$P = \frac{2\pi R}{v} = \frac{2\pi\, 7011 \text{ km}}{7.54 \text{ km/s}}$$

$= 5842$ s $= 5842 \text{ s}\left(\dfrac{1 \text{ min}}{60 \text{ s}}\right)$

$= 97$ min $= 97 \text{ min}\left(\dfrac{1 \text{ hr}}{60 \text{ min}}\right)$ $=1.62$ hr

D. $\boxed{\text{Satellite period} = P = 5842 \text{ s} = 97 \text{ min} = 1.62 \text{ hr}}$

5. Ping pong balls on swimming pool
(Name and Date)

A. DRAWING

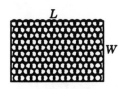
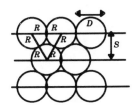

B. DEFINITIONS

Width of pool	$= W$	$= 5$	m
Length of pool	$= L$	$= 15$	m
Diameter of one ball	$= D$	$= 2.5$	cm
Radius of one ball	$= R$		m
Cross-section area of one ball	$= A_1$		m^2
Spacing of ball centers	$= S$		m
Number of balls per meter of width	$= N_W$		
Number of balls per meter of length	$= N_L$		
Total number of balls	$= N$		
Total cross-sectional area of all balls	$= A_b$		m^2
Surface area of pool	$= A_p$		m^2
Fraction of surface covered by balls	$= F$		

C. UNIT CONVERSION

$$D \quad = 2.5 \text{ cm} \quad = 2.5 \text{ cm}\left(\frac{1 \text{ m}}{100 \text{ cm}}\right) \quad = 0.025 \text{ m}$$

Find ball radius

$$R \quad = D/2 \qquad = 0.025 \text{ m} /2 \qquad = 0.0125 \text{ m}$$

Find spacing between rows of balls

$$S \quad = \sqrt{(2R)^2 - R^2} \quad = \sqrt{4R^2 - R^2} \qquad = \sqrt{3R^2}$$
$$= \sqrt{3}\, R \qquad\quad = 1.732\, R \qquad\quad = 1.732\,(0.0125)$$
$$= 0.02165 \text{ m}$$

Find number of balls along length of pool

$$N_l \quad = \frac{L}{D} \qquad\qquad = \frac{15 \text{ m}}{0.025 \text{ m}} \qquad = 600 \text{ balls}$$

Find number of rows along width of pool

$$N_W \quad = \frac{W}{S} \qquad\qquad = \frac{5 \text{ m}}{0.02165} \qquad = 230.95 \quad = 230 \text{ balls}$$

Find total number of balls

$$N \quad = N_L N_W \qquad\qquad = 600 \ (230) \qquad = 138\ 000 \text{ balls}$$

D. $\boxed{\text{Total number of balls} = N = 138\ 000}$

Find cross section area of one ball

$$A_1 \quad = \pi R^2 \qquad\qquad = \pi \ (0.0125)^2 \qquad = 4.91 \text{x} 10^{-4} \text{ m}^2$$

Find total cross sectional area of all balls

$$A_b \quad = N A_1 \qquad\qquad = 138\ 000 \ (4.91 \text{x} 10^{-4} \text{ m}^2) \qquad = 67.76 \text{ m}^2$$

Find total surface area of pool

$$A_p \quad = LW \qquad\qquad = 15 \text{ m} \ (5 \text{ m}) \qquad = 75 \text{ m}^2$$

Find fraction of pool surface covered by balls

$$F \quad = \frac{A_b}{A_p} \qquad\qquad = \frac{67.76 \text{ m}^2}{75 \text{ m}^2} \qquad = 0.90$$

Check algebraically

$$F \quad = \frac{A_b}{A_p} \qquad = \frac{A_1 N}{LW} \qquad = \frac{\pi R^2 N_L N_W}{LW}$$

$$= \frac{\pi R^2 (L/D)(W/S)}{LW} \quad = \frac{\pi R^2}{SD} \quad = \frac{\pi R^2}{\sqrt{3}R \ 2R} \quad = \frac{\pi}{2\sqrt{3}} \quad = 0.907$$

D. $\boxed{\text{Fraction of surface covered by balls} = F = 0.91}$

The fraction covered doesn't depend on the size of the balls or the size of the pool as long as the balls are much smaller than the pool.

7. Memory refresh
(Your name and date)

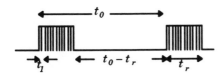

DEFINITIONS

Total number of bits	$= N$	$= 1\ 000\ 000$	
Time between refreshes	$= t_0$	$= 5$	ms
Time to refresh one batch	$= t_1$	$= 50$	ns
Number of bits in one batch	$= N_1$	$= 1\ 000$	
Number of batches	$= N_b$		
Total refresh time	$= t_r$		s
Total useful (non-refresh) time	$= t_u$		s
Fraction of useful time	$= F$		

UNIT CONVERSION

$$t_0 \quad = 5 \text{ ms} \quad = 5 \text{ ms} \left(\frac{10^{-3} \text{ s}}{1 \text{ ms}}\right) \quad = 5 \times 10^{-3} \text{ s}$$

$$t_1 \quad = 50 \text{ ns} \quad = 50 \text{ ns} \left(\frac{10^{-9} \text{ s}}{1 \text{ ns}}\right) \quad = 5 \times 10^{-8} \text{ s}$$

Find number of batches

$$N_b \quad = N/N_1 \quad = 1\ 000\ 000/1000 \quad = 1000$$

Find total refresh time

$$t_r \quad = N_b t_1 \quad = 1\ 000\ (5 \times 10^{-8} \text{s}) \quad = 5 \times 10^{-5} \text{ s}$$

Find total useful time

$$t_u \quad = t_0 - t_r \quad = 5 \times 10^{-3} \text{ s} - 5 \times 10^{-5} \text{ s} \quad = 4.95 \times 10^{-3} \text{ s}$$

Find fraction of time computer is available

$$F \quad = \frac{t_u}{t_0} \quad = \frac{4.95 \times 10^{-3}}{5 \times 10^{-3}} \quad = 0.99$$

Fraction of time memory is available $= F = 0.99$

9. Satellite Height.
(Your name and date)

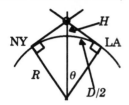

DEFINITIONS

Distance from LA to NY	$= D$		mi
Radius of earth	$= R$	$= 6.37 \times 10^6$	m
Angle at center of earth between vector to satellite and vector to LA	$= \theta$		radians
Height of satellite	$= H$		m

DATA

D = 2794 mi   from Rand McNally Road Atlas of US, Canada, and Mexico. 1983, pg. 105

D = 2794 mi = 2794 mi $\left(\dfrac{1609.3 \text{ m}}{1 \text{ mi}}\right)$ = 4.496×10^6 m

Find angle at center of earth

$\theta \quad = \dfrac{D/2}{R} = \dfrac{4.496 \times 10^6}{2(6.37 \times 10^6)} = 0.3529$ radians

Find height of satellite using right triangle

$\cos\theta \quad = \dfrac{R}{R+H} \quad = \dfrac{1}{1+H/R}$

$(1+H/R)\cos\theta = 1$

$1 + H/R \quad = \dfrac{1}{\cos\theta}$

$\dfrac{H}{R} \quad = \dfrac{1}{\cos\theta} - 1$

$\quad = \dfrac{1}{\cos 0.3529 \text{ rad}} - 1 \quad = \dfrac{1}{0.9384} - 1$

$\quad = 1.06567 - 1 = 0.06567$

$H \quad = 0.06567\ R \quad\quad = 0.06567(6.37 \times 10^6 \text{ m})$

$\quad = 4.18 \times 10^5$ m $\quad\quad = 418$ km

$\boxed{\text{Height of satellite} = H = 418 \text{ km}}$

11. Travelling Piston
(Your name and date)

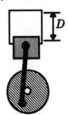

DEFINITIONS

Distance piston travels in 1/2 revolution	$= D$	$= 60$	mm
Speed of car	$= v$	$= 55$	mi/hr
Engine revolutions per minute at 55 mph	$= f$	$= 2600$	min^{-1}
Distance car moves	$= x_c$	$= 1$	m
Distance piston moves	$= x_p$		m
Number of engine revolutions for 1 m of car movement	$= N$		

UNIT CONVERSION

$$D \quad = 60 \text{ mm} \quad = 60 \text{ mm} \left(\frac{1 \text{ m}}{1000 \text{ mm}}\right) \quad = 0.06 \text{ m}$$

$$f \quad = 2600 \frac{\text{rev}}{\text{min}} = 2600 \frac{\text{rev}}{\text{min}} \left(\frac{1 \text{ min}}{60 \text{ s}}\right) \quad = 43.33 \text{ rev/s}$$

$$v \quad = 55 \frac{\text{mi}}{\text{hr}} = 55 \frac{\text{mi}}{\text{hr}} \left(\frac{1609.3 \text{ m}}{1 \text{ mi}}\right)\left(\frac{1 \text{ hr}}{3600 \text{ s}}\right) = 24.59 \text{ m/s}$$

Find the number of engine revolutions for 1 m of car movement (in 1 s the engine makes f revolutions and the car moves v meters)

$$N \quad = \frac{f}{v} \quad = \frac{43.33 \text{ rev/s}}{24.59 \text{ m/s}} \quad = 1.762 \text{ rev/m}$$

Find the distance the piston moves when the car moves 1 m

$$x_p \quad = 2DNx_c \quad = 2 \ (0.06 \text{ m})(1.762 \text{ m}^{-1})(1 \text{ m}) \quad = 0.21 \text{ m}$$

$$\boxed{\text{Distance piston moves} = x_p = 0.21 \text{ m}}$$

13. Pulsed Laser

(Your name and date)

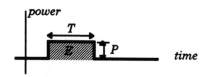

DEFINITIONS

Length of pulse	$= T$	$= 2 \times 10^{-8}$	s
Electrical energy in pulse	$= E_e$	$= 1.25 \times 10^4$	J
Light energy in pulse	$= E_l$		J
Efficiency	$= e$	$= 1\%$	
Power during pulse	$= P$		W

UNIT CONVERSION

$$e \quad = 1\% \qquad = 1\% \left(\frac{1}{100\%} \right) \qquad = 0.01$$

Find light energy in pulse

$$E_l \quad = eE_e$$

Find power during pulse

$$P \quad = \frac{E_l}{T} \quad = \frac{eE_e}{T} \quad = \frac{0.01\,(1.25 \times 10^4)}{2 \times 10^{-8}} \quad = 6.25 \times 10^9 \text{ W}$$

Power during pulse $= P = 6.25 \times 10^9$ W

15. Passing cars
(Your name and date)

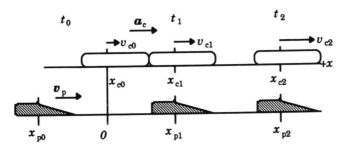

DEFINITIONS

Acceleration of Corvette	$= a_c$	$= 0.8$	g
Acceleration of Porsche	$= a_p$	$= 0$	m/s^2
Velocity of Porsche	$= v_p$	$= 90$	km/hr
Acceleration of gravity	$= g$	$= 9.81$	m/s^2

At $t=0$

Position of Corvette	$= x_{c0}$	$= 0$	m
Velocity of Corvette	$= v_{c0}$	$= 0$	
Position of Porsche	$= x_{p0}$	$= -30$	m

At $t = t_1$ (P passes C)

Time	$= t_1$	s
Position of Corvette	$= x_{c1}$	m
Position of Porsche	$= x_{p1}$	m
Velocity of Corvette	$= v_{c1}$	m/s

At $t=t_2$ (C passes P)

Time	$= t_2$	s
Position of Corvette	$= x_{c2}$	m
Position of Porsche	$= x_{p2}$	m
Velocity of Corvette	$= v_{c2}$	m/s

PRELIMINARY EQUATIONS

$$a_c = 0.8\, g = 0.8\,(\,9.81\ m/s^2) = 7.848\ m/s^2$$

$$v_p = 90\ km/hr = 90\left(\frac{km}{hr}\right)\left(\frac{1000m}{1\ km}\right)\left(\frac{1\ hr}{3600\ s}\right) = 25\ m/s$$

For motion with constant acceleration

$$a = constant$$
$$v = v_0 + at$$
$$x = x_0 + v_0 t + \tfrac{1}{2}at^2$$

When the Porsche passes the Corvette, $t=t_1$, and $x_{p1}=x_{c1}$

For the Corvette

$$x_{c1} = x_{c0} + v_{c0}t_1 + \tfrac{1}{2}a_c t_1^2$$
$$= 0 + 0 + \tfrac{1}{2}a_c t_1^2$$

For the Porsche

$$x_{p1} = x_{p0} + v_p t_1 + \tfrac{1}{2}a_p t_1^2$$
$$= x_{p0} + v_p t_1 + 0$$

Equate x_{c1} and x_{p1}

$$x_{c1} = x_{p1}$$
$$\tfrac{1}{2}a_c t_1^2 = x_{p0} + v_p t_1$$

Solve for t_1

$$\tfrac{1}{2}a_c t_1^2 - v_p t_1 - x_{p0} = 0$$

This looks like

$$at^2 + bt + c = 0$$

So

$$t = \frac{-b \pm \sqrt{b^2 - 4a\,c}}{2a}$$

And

$$t_1 = \frac{+v_p \pm \sqrt{v_p^2 - 4\left(\tfrac{1}{2}a_c\right)(-x_{p0})}}{a_c}$$

$$= \frac{v_p \pm \sqrt{v_p^2 + 2a_c x_{p0}}}{a_c}$$

$$= \frac{25 \text{ m/s} \pm \sqrt{(25 \text{ m/s})^2 + 2\,(7.848 \text{ m/s}^2)\,(-30 \text{ m})}}{7.848 \text{ m/s}^2}$$

$$= \frac{25 \pm \sqrt{625 - 470.9}}{7.848} = \frac{25 \pm \sqrt{154.1}}{7.848} = \frac{25 \pm 12.4}{7.848}$$

To find t_1, the smaller root

$$t_1 = \frac{25 - 12.4}{7.848} = \frac{12.6}{7.848} = 1.60 \text{ s}$$

To find t_2, the larger root

$$t_2 = \frac{25 + 12.4}{7.848} = \frac{37.4}{7.848} = 4.77 \text{ s}$$

To find x_1, the position at which the Porsche passes the Corvette

x_{c1} $= \frac{1}{2}a_c t_1^2$ $= \frac{1}{2}$ (7.848 m/s^2) (1.60 s)^2 $= 10.05$ m

Check using position of Porsche at t_1

x_{p1} $= x_{p0} + v_p t_1$ $= -30$ m + 25 m/s (1.60 s) $= 10.0$ m

to find x_2, the position at which the Corvette passes the Porsche

x_{c2} $= \frac{1}{2}a_c t_2^2$ $= \frac{1}{2}$ (7.848 m/s^2) (4.77 s)^2 $= 89.28$ m

Check using position of Porsche at t_2

x_{p2} $= x_{p0} + v_p t_2$ $= -30$ m + 25 m/s (4.77 s) $= 89.25$ m

Position at which Porsche passes Corvette = x_1 = 10.0 m

Position at which Corvette passes Porsche = x_2 = 89.3 m

What is the time at which the Corvette gets a speeding ticket?

17. Human-powered plane.
(Your name and date)

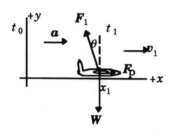

DEFINITIONS

Weight of plane	$= W$	$= 220$	lb
Horizontal acceleration of plane	$= a_x$	$= 0.5$	m/s
Vertical acceleration of plane	$= a_y$	$= 0$	m/s^2
Constant for force on wings	$= K$	$= 30$	N-s^2/m^2

At $t = 0$

Horizontal position of plane	$= x_0$	$= 0$	m
Velocity of plane	$= v_0$	$= 0$	m/s

At t_1 (when plane just lifts off ground)

Time	$= t_1$		sec
Position	$= x_1$	$= 33$	m
Velocity	$= v_1$		m/s
Force on wings	$= F_1$		N
Upward component of force	$= F_y$		N
Angle of force from vertical	$= \theta$		°

PRELIMINARY EQUATIONS

$$W = 220 \text{ lb} = 220 \text{ lb} \left(\frac{4.448 \text{ N}}{1 \text{ lb}}\right) = 978.6 \text{ N}$$

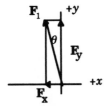

$$F_y = F_1 \cos \theta$$
$$F_x = F_1 \sin \theta$$

At takeoff time t_1

$$F_y = W$$
$$F_1 = K v_1^2$$
$$\frac{F_y}{\cos\theta} = K v_1^2$$
$$\cos\theta = \frac{F_y}{K v_1^2} = \frac{W}{K v_1^2}$$

To find v_1

For motion with constant acceleration

$$a = \text{constant}$$
$$v = v_0 + at$$
$$x = x_0 + v_0 t + \tfrac{1}{2} a t^2$$

At t_1, when the plane leaves the ground

$$x_1 = 0 + 0 + \tfrac{1}{2} a_x t_1^2$$
$$t_1^2 = \frac{2x_1}{a_x}$$

Find v_1 in terms of t_1

$$v_1 = 0 + a_x t_1$$

But we need v_1^2

$$v_1^2 = a_x^2 t_1^2 = a_x^2 \left(\frac{2x_1}{a_x} \right) = 2 a_x x_1$$

Back to $\cos\theta$

$$\cos\theta = \frac{W}{K v_1^2} = \frac{W}{K (2 a_x x_1)}$$
$$= \frac{978.6 \text{ N}}{(30 \text{ N-s}^2/\text{m}^2) \, 2 \, (0.5 \text{ m/s}^2) \, (33 \text{ m})} = 0.9885$$
$$\theta = 8.7\,°$$

Angle of force from vertical = $\theta = 8.7°$

108

19. Earthly Speed
(Your name and date)

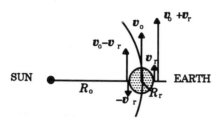

DEFINITIONS

Radius of earth's orbit about sun	$= R_o$		m
Radius of earth	$= R_r$		m
Period of one orbit	$= T_o$	$= 365$	days
Period of one rotation	$= T_r$	$= 24$	hours
Velocity of center of earth in orbit	$= v_o$		m/s
Velocity at equator from rotation	$= v_r$		m/s
Latitude	$= \theta$		°
Velocity at latitude θ from rotation	$= v$		m/s
Maximum velocity at latitude θ	$= v_{max}$		m/s
Minimum velocity at latitude θ	$= v_{min}$		m/s

DATA

R_o $= (91.4 \text{ to } 94.6) \times 10^6$ mi   The World Almanac, World Almanac Co., 1990, pg. 246

R_r $= 3963$ mi   World Almanac, pg. 246

For New York:

θ $= 40°45'$ World Almanac, pg 265

UNIT CONVERSIONS

$$R_o = \left(\frac{91.4+94.6}{2}\right) \times 10^6 \text{ mi} = 93 \times 10^6 \text{ mi}$$

$$= 93 \times 10^6 \text{ mi} \left(\frac{1609.3 \text{ m}}{1 \text{ mi}}\right) = 1.496 \times 10^{11} \text{ m}$$

$$R_r = 3963 \text{ mi} \left(\frac{1609.3 \text{ m}}{1 \text{ mi}}\right) = 6.378 \times 10^6 \text{ m}$$

$$T_o = 365 \text{ d} \left(\frac{24 \text{ h}}{1 \text{ d}}\right)\left(\frac{3600 \text{ s}}{1 \text{ h}}\right) = 3.154 \times 10^7 \text{ s}$$

$$T_r = 24 \text{ h} \left(\frac{3600 \text{ s}}{1 \text{ h}}\right) = 8.64 \times 10^4 \text{ s}$$

Find the orbital velocity
$$v_o = \frac{2\pi R_o}{T_o} = \frac{2\pi(1.496 \times 10^{11}\text{m})}{3.154 \times 10^7 \text{ s}} = 2.980 \times 10^4 \text{ m/s}$$

Find the rotational velocity at the equator
$$v_r = \frac{2\pi R_r}{T_r} = \frac{2\pi(6.378 \times 10^6 \text{m})}{8.64 \times 10^4 \text{ s}} = 4.64 \times 10^2 \text{ m/s}$$

To calculate the distance from the earth's axis to its surface at a latitude of θ
$$R = R_r\cos\theta$$

To calculate the velocity at a latitude of θ caused by rotation
$$v = \frac{2\pi R}{T_r} = \frac{2\pi R_r\cos\theta}{T_r}$$

At $\theta=40°45$
$$v = \frac{2\pi(6.378 \times 10^6 \text{ m}) (\cos 40°45')}{8.64 \times 10^4 \text{ s}}$$
$$= \frac{2\pi(6.378 \times 10^6 \text{ m}) (0.7576)}{8.64 \times 10^4 \text{ s}} = 3.51 \times 10^2 \text{ m/s}$$

To find the maximum velocity
$$v_{max} = v_o + v = 2.980 \times 10^4 + 3.51 \times 10^2 = 3.015 \times 10^4 \text{ m/s}$$
Maximum velocity = v_{max} = 3.02×10^4 m/s

To find the minimum velocity
$$v_{min} = v_o - v = 2.980 \times 10^4 - 3.51 \times 10^2 = 2.945 \times 10^4 \text{ m/s}$$
Minimum velocity = v_{min} = 2.95×10^4 m/s

v_o and v are in the same direction when you are on the side of the earth away from the sun. That is, at midnight.

Maximum velocity occurs at midnight

Index